$165.00

Biofluid Mechanics in Cardiovascular Systems

McGraw-Hill's Biomedical Engineering Series

HJORTSØ • *Population Balances in Biomedical Engineering: Segregation through the Distribution of Cell States*

PANCHAPAKESAN • *Biomedical Nanotechnology*

PETERS • *Real-Time Biomolecular Simulations*

SARKODIE-GYAN • *Neurorehabilitation Devices: Engineering Design, Measurement, and Control*

WAITE • *Biofluid Mechanics in Cardiovascular Systems*

Biofluid Mechanics in Cardiovascular Systems

Lee Waite

Rose-Hulman Institute of Technology
Terre Haute, Indiana

McGraw-Hill

New York Chicago San Francisco Lisbon London Madrid Mexico City Milan New Delhi San Juan Seoul Singapore Sydney Toronto

The McGraw-Hill Companies

CIP Data is on file with the Library of Congress

1 2 3 4 5 6 7 8 9 0 DOC/DOC 0 1 0 9 8 7 6 5

ISBN 0-07-144788-1

The sponsoring editor for this book was Kenneth P. McCombs and the production supervisor was Richard C. Ruzycka. It was set in Century Schoolbook by International Typesetting and Composition. The art director for the cover was Handel Low.

Printed and bound by RR Donnelley.

This book is printed on recycled, acid-free paper containing a minimum of 50% recycled, de-inked fiber.

Contents

Preface

Biomedical engineering is a discipline that is multidisciplinary by definition. The days when medicine was left to the physicians, and engineering was left to the engineers, seem to have passed us by. I have searched for an undergraduate or graduate-level biomedical fluid mechanics textbook since I began teaching at Rose-Hulman in 1987. I looked for, but never found, a book that combined the physiology of the cardiovascular system with engineering of fluid mechanics and hematology to my satisfaction, so I agreed to write one.

This first attempt is published as a monograph on biomedical fluid mechanics, rather than as a textbook. The book does not include problem sets and a solution manual that traditionally accompany engineering textbooks. This work, however, will also form the basis of the next edition that is planned as a textbook.

Biofluid Mechanics in Cardiovascular Systems begins in Chapter 1 with a review of some of the basics of fluid mechanics that all mechanical or chemical engineers would learn. It continues with two chapters on cardiovascular and pulmonary physiology followed by a chapter describing hematology and blood rheology. These four chapters provide the foundation for the remainder of the book.

My 10-week, graduate-level, biofluid mechanics course forms the basis for the book. The course consists of 40 lectures and covers most, but not all, of the material contained in the book. The course is intended to prepare some students for work in the health care device industry and other for graduate work in biomedical engineering.

In spite of great effort on the part of many proofreaders, mistakes may appear in this book. I welcome suggestions for improvement from all readers, with intent to improve subsequent printings and editions.

Lee Waite

Acknowledgments

I would like to thank my BE525 students from spring term of 2005, who helped to proofread the draft version of *Biofluid Mechanics in Cardiovascular Systems*. Thanks to Adnan Ayub, my student, who worked hard to bring me the errors he found in the initial version. Special thanks to another student, Megan Whitaker, an outstanding artist and exceptional proofreader, who also helped to find errors in my writings so that future readers won't be forced to see them.

I wish to thank Ken McCombs and the editorial and production staff at McGraw-Hill for their assistance.

Thanks to my friend and colleague Dr. Jerry Fine, the unofficial Dean of Mountain Climbing at Rose-Hulman, who is responsible for significant improvements in Chapters 7 to 10 because of his very helpful comments.

Thanks to the faculty in the Applied Biology & Biomedical Engineering Department at Rose-Hulman for their counsel and for putting up with a department head who wrote a book instead of giving full concentration to departmental issues.

In writing this book I am keenly aware of the debt I owe to Don Young, my dissertation advisor at Iowa State, who taught me much of what I know about biomedical fluid mechanics.

Thanks to my adult children Bill and Sarah for their support and for staying out of trouble long enough for me to write a book. They are great kids.

Most of all I would like to thank my colleague, wife, and best friend Gabi Nindl Waite, who put up with me during the long evening hours and long weekends that it took to write this book. Thanks especially for everything she taught me about physiology.

Chapter

1

Review of Basic Fluid Mechanics Concepts

1.1 A Brief History of Biomedical Fluid Mechanics

People have written about circulation for thousands of years. I include here a short history of biomedical fluid mechanics, because I believe it is important to recognize that in all of science and engineering we "stand on the shoulders of giants."[1]

The Yellow Emperor, Huang Ti, lived in China between 475 and 221 B.C. and he wrote one of the first works dealing with circulation. Huang Ti wrote "Internal Classics," in which fundamental theories of Chinese medicine were addressed. Among other topics, Huang Ti wrote about the Yin Yang doctrine and the theory of circulation.

Hippocrates (Fig. 1.1) lived in Greece in 400 B.C., is considered by many to be the father of science-based medicine, and was the first to separate medicine from magic. Hippocrates declared that the human body was integral with nature and was something that should be understood. He founded a medical school on the island of Cos, Greece, and developed the Oath of Medical Ethics. Hippocrates lived until 377 B.C.

Aristotle lived in Greece between 384 and 322 B.C. He wrote that the heart was the focus of the blood vessels, but did not make a distinction between arteries and veins.

[1]"If I have seen further (than others) it is by standing on the shoulders of giants." This quote was written by Isaac Newton in a letter to Robert Hooke, 1675.

Figure 1.1 Hippocrates. (Courtesy of the National Library of Medicine Images from the History of Medicine B029254)

Praxagoras of Cos was a Greek physician and a contemporary of Aristotle. Praxagoras was apparently the first Greek physician to recognize the difference between arteries (carriers of air, he thought) and veins (carriers of blood), and to comment on the pulse.

William Harvey (Fig. 1.2) was born in Folkstone, England, in 1578. He earned a BA degree from Cambridge in 1597 and went on to study medicine in Padua, Italy, where he received his doctorate in 1602. Harvey returned to England to open a medical practice. He married Elizabeth Brown, daughter of the court physician to Queen Elizabeth I and King James I. Harvey eventually became court physician to King James I and King Charles I.

In 1628, Harvey published "An anatomical study of the motion of the heart and of the blood of animals." This was the first publication that claimed that blood is pumped from the heart and recirculated. Up to that point, the common theory of the day was that food was converted to

Figure 1.2 William Harvey. (Courtesy of the National Library of Medicine Images from the History of Medicine B014191)

blood in the liver and then consumed as fuel. To prove that blood was recirculated and not consumed, Harvey showed, by calculation, that blood pumped from the heart in only a few minutes exceeded the total volume of blood contained in the body.

Jean Louis Marie Poiseuille was a French physician and physiologist born in 1797. Poiseuille studied physics and math in Paris. Later he became interested in the flow of human blood in narrow tubes and in 1838, he experimentally derived and later published Poiseuille's law. Poiseuille's law describes the relationship between flow and pressure-gradient in long tubes with constant cross-section. Poiseuille died in Paris in 1869.

Otto Frank was born in Germany in 1865 and he died in 1944. He was educated in Munich, Kiel, Heidelberg, Glasgow, and Strasburg. In 1890 Frank published, "Fundamental form of the arterial pulse," which contained his "Windkessel Theory," of circulation. He became a physician in 1892 in Leipzig and became a professor in Munich in 1895. Frank perfected optical manometers and capsules for the precise measurement of intracardiac pressures and volumes.

1.2 Fluid Characteristics and Viscosity

A fluid is defined as a substance that deforms continuously under application of a shearing stress, regardless of how small the stress. Blood is the primary example of a biological material that behaves as a fluid about which I will write in this text. To study the behavior of materials that act as fluids, it is useful to define a number of fluid parameters. A list of important fluid characteristics would include density, specific weight, specific gravity, and viscosity.

Density is defined as the mass per unit volume of a substance and is given by the Greek character ρ (rho). The SI units for ρ are kg/m^3 and the approximate density for blood is 1060 kg/m^3. Blood is slightly denser than water, and red blood cells in plasma[2] will settle to the bottom of a test tube, over time, due to gravity.

Specific weight is defined as the weight per unit volume of a substance. The SI units for specific weight are N/m^3. Specific gravity, s, is the ratio of the weight of a liquid at a standard reference temperature to the weight of water. For example, the specific weight of mercury, $S_{HG} = 13.6$ at 20 °C. Specific gravity is a unitless parameter.

Density and specific weight are measures of the "heaviness" of a fluid, but two fluids with identical density and specific weight can flow quite differently when subjected to the same forces. You might ask, "What is the additional property that determines the difference in behavior?" That property is viscosity.

1.2.1 Displacement

To understand viscosity, begin by imagining a hypothetical fluid between two parallel plates which are infinite in width and length. See Fig. 1.3.

The bottom plate A is a fixed plate. The upper plate B is a moveable plate, suspended above plate A on the fluid between the two plates. The vertical distance between the two plates is represented by h. A constant force F is applied to the moveable plate B causing it to move along at a constant velocity, V_B, with respect to the fixed plate.

If we replace the fluid between the two plates with a solid, the behavior of the plates would be different. The applied force F would create a displacement, d, a shear stress, τ, in the material, and a shear strain, γ. After a small, finite displacement, motion of the upper plate would cease.

If we then replace the solid between the two plates with a fluid, and reapply force F, the upper plate will move continuously, with a velocity of V_B. This behavior is consistent with the definition of a fluid; a material that deforms continuously under the application of a shearing stress, regardless how small the stress.

---

[2]Plasma has a density very close to that of water.

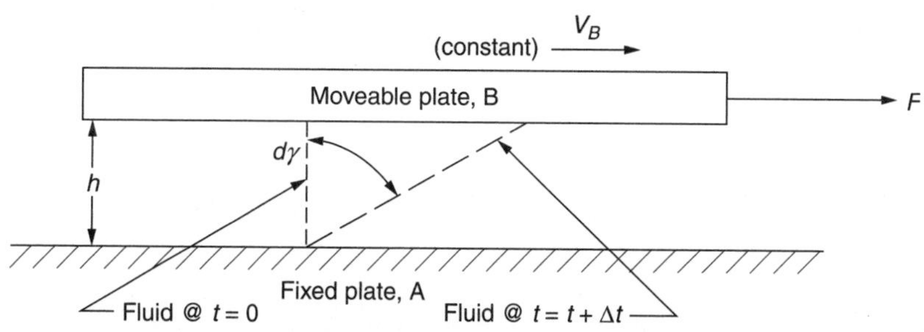

Figure 1.3 Moveable plate suspended over a layer of fluid.

After some finite length of time, dt, a line of fluid that was vertical at time = 0 will move to a new position as shown by the dashed line in Fig. 1.3. The angle between the line of fluid at $t = 0$ and the line of fluid at $t = t + dt$, is defined as the shearing strain. Shearing strain is represented by the Greek character γ (gamma).

The first derivative of the shearing strain with respect to time is known as the rate of shearing strain, $d\gamma/dt$. For small displacements, $\tan(d\gamma)$ is approximately equal to $d\gamma$. The tangent of the angle of shearing strain can also be represented as follows:

$$\tan(d\gamma) = \frac{\text{opposite}}{\text{adjacent}} = \frac{V_B dt}{h}$$

Therefore, the rate of shearing strain, $d\gamma/dt$ can be written:

$$d\gamma/dt = V_B/h$$

The rate of shearing strain may also be written as $\dot{\gamma}$, and has the units of 1/s.

Velocity. The fluid that touches plate A has zero velocity. The fluid that touches plate B moves with the same velocity, V_B, as plate B. That is, the molecules of fluid adhere to the plate and do not slide along its surface. This is known as the no-slip condition. The no-slip condition is important in fluid mechanics. All fluids, including both gasses and liquids, satisfy this condition.

Let the distance from the fixed plate to some arbitrary point above the plate be y. The velocity of the fluid between the plates, V, is a function of the distance above the fixed plate A. More succinctly written:

$$V = V(y)$$

The velocity of the fluid at any point between the plates varies linearly between $V = 0$ and $V = V_B$. See Fig. 1.4.

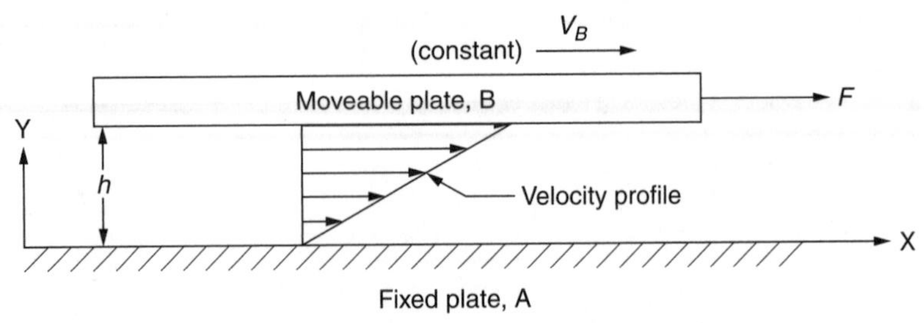

Figure 1.4 Velocity profile in a fluid between two parallel plates.

Let us define the velocity gradient as the change in fluid velocity with respect to y

$$\text{Velocity gradient} \equiv dV/dy$$

The velocity profile is a graphical representation of the velocity gradient. See Fig. 1.4. For a linearly varying velocity profile like that shown in Fig. 1.4 the velocity gradient can also be written as

$$\text{Velocity gradient} \equiv V_B/h$$

1.2.2 Shear stress

In cardiovascular fluid mechanics, shear stress is a particularly important concept. Blood is a living fluid, and if the forces applied to the fluid are sufficient, the resulting shearing stress can cause red blood cells to be destroyed. On the other hand, studies indicate a role of shear stress in modulating atherosclerotic plaques. The relationship between shear stress and atherosclerosis (arterial disease) is studied much but not very well understood.

Figure 1.5 represents shear stress on an element of fluid at some arbitrary point between the plates in Figs. 1.3 and 1.4. The shear stress on the top of the element results in a force that pulls the element "downstream." The shear stress at the bottom of the element resists that movement.

Since the fluid element shown will be moving at a constant velocity, and not rotating, the shear stress, τ' on the element must be the same as the shear stress τ. Therefore,

$$\frac{d\tau}{dy} = 0 \qquad \text{and} \qquad \tau_A = \tau_B = \tau_{\text{wall.}}$$

Physically, the shearing stress at the wall may also be represented by:

$$\tau_A = \tau_B = \frac{\text{force}}{\text{plate area}}$$

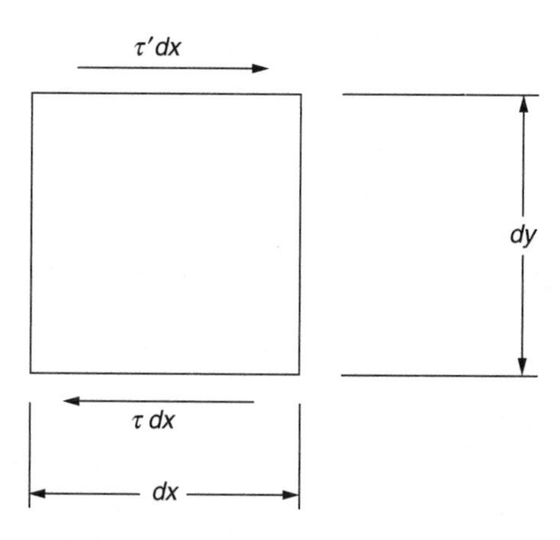

Figure 1.5 Shear stresses on an element of fluid.

The shear stress on a fluid is related to the rate of shearing strain. If a very large force is applied in moving plate B, a relatively higher velocity, higher rate of shearing strain, and a higher stress will result. In fact, the relationship between shearing stress and rate of shearing strain is determined by the fluid property known as viscosity.

Viscosity. A common way to visualize material properties in fluids is by making a plot of shearing stress as a function of the rate of shearing strain. For the plot shown in Fig. 1.6, shearing stress is represented by the Greek character τ, and the rate of shearing strain is represented by $\dot{\gamma}$.

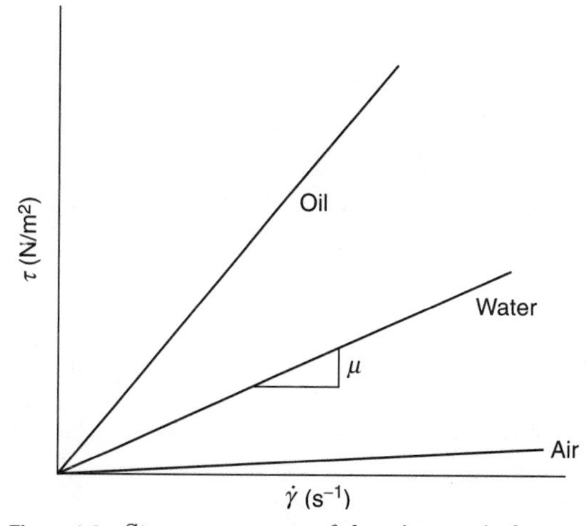

Figure 1.6 Stress versus rate of shearing strain for various fluids.

The material property that is represented by the slope of the stress-shearing rate curve is known as viscosity and is represented by the Greek letter μ, (mu). Viscosity is also sometimes referred to by the name absolute viscosity or dynamic viscosity.

For common fluids like oil, water, and air, viscosity does not vary with shearing rate. Fluids with constant viscosity are known as newtonian fluids. For newtonian fluids, shear/stress rate of shearing strain may be related in the following equation:

$$\tau = \mu\dot{\gamma}$$

where τ = shear stress
μ = viscosity
$\dot{\gamma}$ = the rate of shearing strain

For non-newtonian fluids τ and $\dot{\gamma}$ are not linearly related. For those fluids, viscosity can change as a function of the shear rate (rate of shearing strain). Blood is an important example of a non-newtonian fluid. Later in this book, we will investigate the condition under which blood behaves as, and may be considered, a newtonian fluid.

Shear stress and shear rate are not linearly related for non-newtonian fluids. Therefore, the slope of the shear stress/shear rate curve is not constant. However, we can still talk about viscosity if we define the apparent viscosity as the instantaneous slope of the shear stress/shear rate curve. See Fig. 1.7.

Shear thinning fluids are non-newtonian fluids whose apparent viscosity decreases as shear rate increases. Latex paint is a good example of

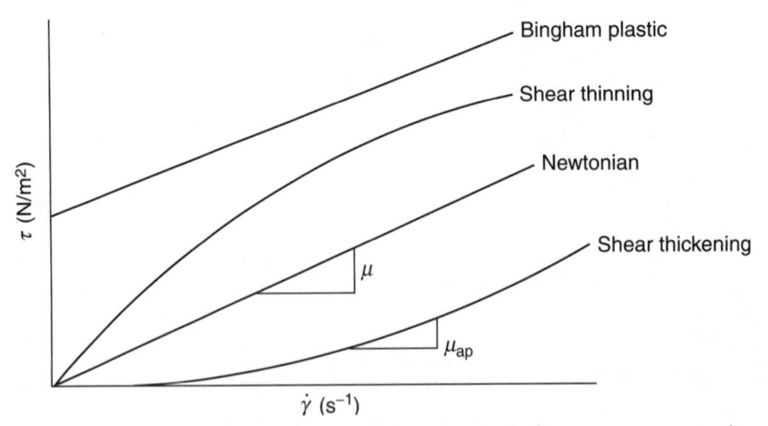

Figure 1.7 Shear stresses versus rate of shearing strain for some non-newtonian fluids.

a shear thinning fluid. It is a positive characteristic of paint that the viscosity is low when one is painting, but that the viscosity becomes higher and the paint sticks to the surface better when no shearing force is present. At low shear rates, blood is also a shear thinning fluid. However when the shear rate increases above 100 s^{-1}, blood behaves as a newtonian fluid.

Shear thickening fluids are non-newtonian fluids whose apparent viscosity increases when the shear rate increases. Quicksand is a good example of a shear thickening fluid. If one tries to move slowly in quicksand, then the viscosity is low and movement is relatively easy. If one tries to struggle and move quickly, then the viscosity increases. A mixture of cornstarch and water also forms a shear thickening non-newtonian fluid.

A Bingham plastic is not a fluid but also not a solid. A Bingham plastic can withstand a finite shear load but can flow like a fluid when that shear stress is exceeded. Toothpaste and mayonnaise are examples of Bingham plastics. Blood is also a Bingham plastic and behaves as a solid at shear rates very close to zero. The yield stress for blood is very small, approximately 0.005 to 0.01 N/m^2.

Kinematic viscosity is another fluid property that is used to characterize flow. Kinematic viscosity is the ratio of absolute viscosity to fluid density. Kinematic viscosity is represented by the Greek character ν (nu). Kinematic viscosity may be represented by the equation:

$$\nu = \frac{\mu}{\rho}$$

where μ is absolute viscosity and ρ is fluid density.

The SI units for absolute viscosity are Ns/m^2. The SI units for kinematic viscosity are m^2/s.

1.3 Fundamental Method for Measuring Viscosity

A fundamental method for measuring viscosity involves a viscometer made from concentric cylinders. See Fig. 1.8. The fluid for which the viscosity is to be measured is placed between the two cylinders. The torque generated on the inside fixed cylinder by the outer rotating cylinder is measured through a torque measuring shaft. The force required to cause the cylinder to spin and the velocity at which it spins are also measured. Then the viscosity may be calculated in the following way:

The shear stress in the fluid is equal to the force, F applied to the outer cylinder divided by the surface area of the internal cylinder, A.

$$\tau = \frac{F}{A}$$

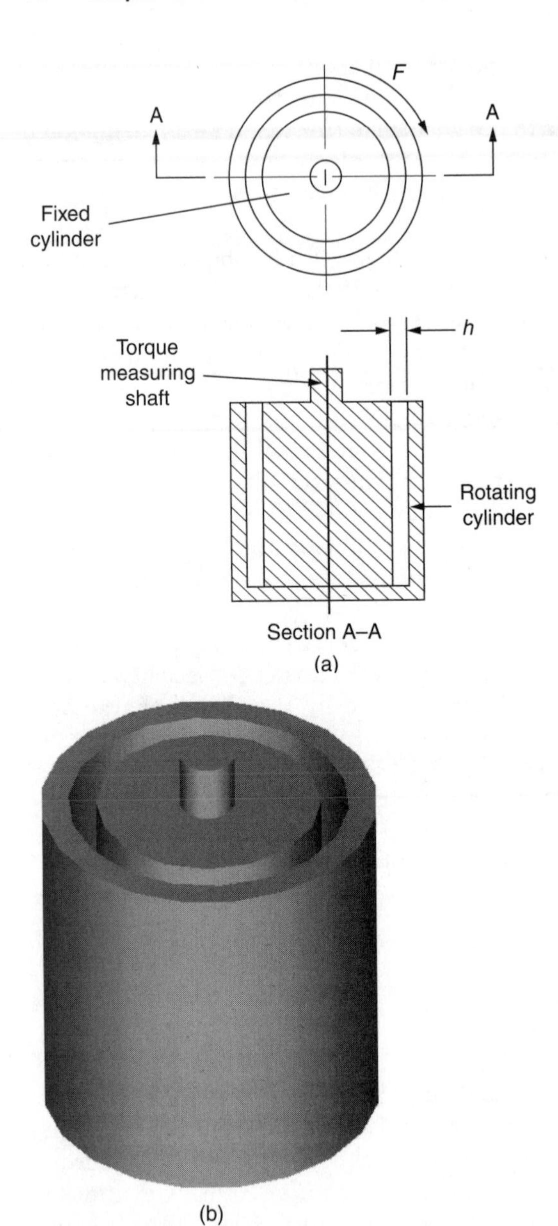

Figure 1.8 (a) Cross-section of a rotating cylinder viscometer. (b) Rotating cylinder viscometer.

The shear rate, $\dot{\gamma}$, for the fluid in the gap between the cylinders may also be calculated from the velocity of the cylinder, V, and the gap width, h:

$$\dot{\gamma} = \frac{V}{h}$$

From the shear stress and the shear rate, the viscosity and/or kinematic velocity may be obtained:

$$\mu = \frac{\tau}{\dot{\gamma}} \qquad \text{and} \qquad \nu = \frac{\mu}{\rho}$$

where μ = the viscosity,
ν = the kinematic viscosity
ρ = the density.

A typical value for blood viscosity in humans is 0.0035 Ns/m^2 or 0.035 poise (P) or 3.5 cP. A poise is a dyne cm/s^2. A poise is also equal to 0.1 Ns/m^2. Another useful pressure unit conversion is that 1 mmHg = 133.3 N/m^2.

1.4 Introduction to Pipe Flow

An eulerian description of flow is one in which a field concept is used. Descriptions which make use of velocity fields and flow fields are eulerian descriptions. Another type of flow description is a lagrangian description. In the lagrangian description, particles are tagged and the paths of those particles are followed. An eulerian description of goose migration could involve you or me sitting on the shore of Lake Erie and counting the number of geese that fly over in an hour. In a lagrangian description of the same migration we might capture and band a single goose with a radio transmitter and study the path of the goose, including its position and velocity as a function of time.

Consider an eulerian description of flow through a constant cross-section pipe as shown in Fig. 1.9. In the figure, the fluid velocity is shown across the pipe cross-section and at various points along the length of the pipe. In the entrance region the flow begins with a relatively flat velocity profile and then develops an increasingly parabolic flow profile as distance x along the pipe increases. Once the flow profile becomes constant and no longer changes with increasing x the velocity profile does not change. The region in which the velocity profile is constant is known as the region of fully developed flow.

The pressure gradient is the derivative of pressure with respect to distance. Mathematically the pressure gradient is written as dP/dx, where P is the pressure inside the pipe at some point and x is the distance along the pipe. In the region of fully developed flow, the pressure gradient,

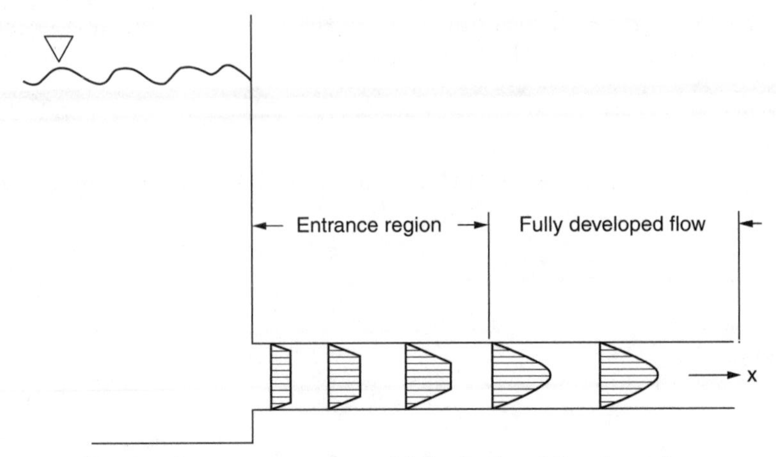

Figure 1.9 Showing the entrance region and fully developed flow in a tube.

dP/dx is constant. On the other hand, the pressure gradient in the entrance region varies with position x, as shown in the plot below in Fig. 1.10. The slope of the plot is the pressure gradient.

1.4.1 Reynolds number

Osborne Reynolds was a British engineer, born in 1842 in Belfast. In 1895, he published the paper, "On the dynamical theory of incompressible viscous fluids and the determination of the criterion." The paper was a landmark contribution to the development of fluid mechanics and the crowning achievement of Reynolds' career. He was the first professor of engineering at the University of Manchester, England.

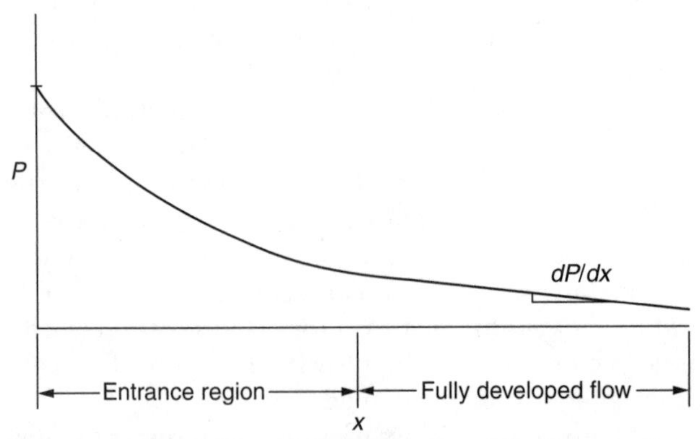

Figure 1.10 Showing pressure as a function of the distance along a pipe. Note that the pressure gradient, dP/dx is constant for fully developed flow.

The Reynolds number is a dimensionless parameter named after Dr. Reynolds. The number is defined as

$$\mathrm{Re} = \frac{\rho V D}{\mu}$$

where ρ = fluid density in kg/m^3
V = fluid velocity in m/s
D = pipe diameter in m
μ = fluid viscosity in Ns/m^2

Unless otherwise specified, this V will be considered to be the average velocity across the pipe cross-section. Physically the Reynolds number represents the ratio of inertial forces to viscous forces.

The Reynolds number (Re) helps us to predict the transition between laminar and turbulent flow. Laminar flow is highly organized flow along streamlines. When velocity increases, flow can become disorganized with a random 3D motion superimposed on the average flow velocity. This is known as turbulent flow. Laminar flow occurs in flow environments where Re < 2000. Turbulent flow is present in circumstances under which Re > 4000. The range of 2000 < Re < 4000 is known as the transition range.

The Reynolds number is also useful for predicting entrance length in pipe flow. I will write entrance length as, X_E. The ratio of entrance length to pipe diameter for laminar pipe flow is given as

$$\frac{X_E}{D} \cong 0.06\ \mathrm{Re}$$

Consider the following example: If Re = 300, then Xe = 18 D and an entrance length equal to 18 pipe diameters is required for fully developed flow. In the human cardiovascular system, it is not common to see fully developed flow in arteries. Typically the vessels continually branch, with the distance between branches not often being greater than 18 diameters.

Although most blood flow in humans is laminar, having a Re of 300 or less, it is possible for turbulence to occur at very high flow rates in the descending aorta, for example in highly conditioned athletes. Turbulence is also common in pathological conditions such as heart murmurs and stenotic heart valves.

Stenotic comes from the Greek, "stenos" meaning narrow. Stenotic means narrowed and a stenotic heart valve is one in which the narrowing of the valve is the result of a plaque formation on the valve.

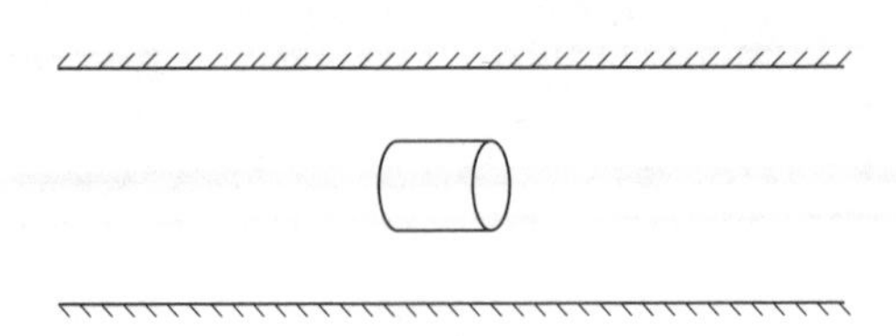

Figure 1.11 An element of fluid in pipe flow.

1.4.2 Poiseuille's law

In 1838 Jean-Marie Poiseuille empirically derived this law, which is also known as the Hagen-Poiseuille law, for Gotthilf Heinrich Ludwig Hagen for his experiments in 1839. This law describes steady, laminar, incompressible, viscous flow of a newtonian fluid in a rigid, cylindrical tube of constant cross-section. Poiseuille published the law in 1840.

Consider a cylinder of fluid in a region of fully developed flow. See Fig. 1.11.

Next I draw a free-body diagram of the cylinder of fluid and sum the forces acting on that cylinder. See Fig. 1.12 for a free-body diagram.

First, we will need to make some assumptions. Assume that the flow is steady. This means that the flow is not changing with time; that the derivative of flow rate with respect to time is equal to zero.

$$\frac{dQ}{dt} = 0$$

Second, assume that the flow is through a long tube with a constant cross-section. This flow condition is known as uniform flow. For steady flows in long tubes with a constant cross-section, the flow is fully developed and therefore, the pressure gradient, dP/dx is constant.

The third assumption is that the flow is newtonian. newtonian flow is flow in which the shearing stress, τ, in the fluid is constant. In other words, the viscosity is constant with respect to the shear rate, $\dot{\gamma}$, and the whole process is carried out at a constant temperature.

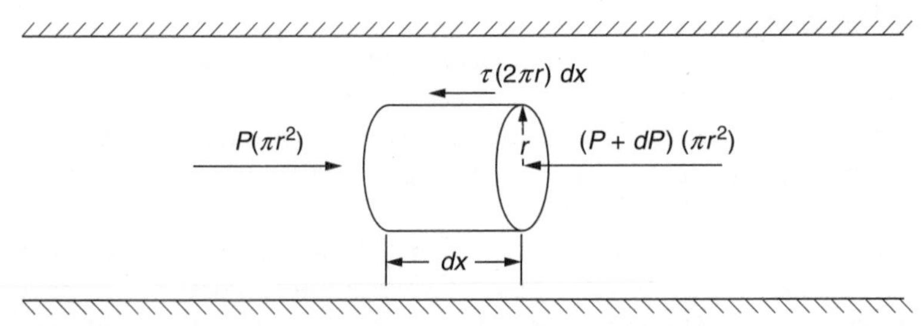

Figure 1.12 Free-body diagram of the forces acting on an element of fluid in pipe flow.

Now let the x direction be the axial direction of the pipe with downstream (to the right) being positive. If the flow is unchanging with time, then the sum of forces in the x direction is zero and Eq. (1.1) is written:

$$P(\pi r^2) - (P + dP)(\pi r^2) - \tau 2\pi rdx = 0 \qquad (1.1)$$

Solve Eq. (1.1) and the result is

$$-dP(\pi r^2) = 2\pi r\tau\, dx \qquad (1.2)$$

The result from a simple force balance is shear stress as a function of pressure gradient, dP/dx and radial position, r:

$$\tau = -\frac{r}{2}\frac{dP}{dx} \qquad (1.3)$$

and

$$\tau_{\text{wall}} = \frac{-R_{\text{tube}}}{2}\frac{dP}{dx} \qquad (1.4)$$

Recall from the definition of viscosity that the shear stress is also related to the shear rate in the following way:

$$\tau = -\mu\, dV/dr \qquad (1.5)$$

Now if we solve Eqs. (1.4) and (1.5) together, the resulting expression is an equation relating shear rate to pressure gradient:

$$\frac{-r}{2}\frac{dP}{dx} = -\mu\frac{dV}{dr} \qquad (1.6)$$

Then, separating the variables, this expression produces a differential equation with the variables velocity, V, and radius, r:

$$dV = \frac{1}{2\mu}\frac{dP}{dx} rdr \qquad (1.7)$$

The next step in the analysis is to solve the differential equation above, producing Eq. (1.8), which gives the velocity of each point in the tube as a function of the radius, r:

$$V = \frac{1}{2\mu}\frac{dP}{dx}\frac{r^2}{r} + C_1 \qquad (1.8)$$

So far, in this analysis, we have made three assumptions. First, steady flow ($dQ/dt = 0$); second, fully developed flow in a constant cross-section tube (dP/dx is constant); and third, viscosity (μ), is constant. Now we need to make assumption four, which is the no slip condition. This means V at the wall is zero when r equals the radius of the tube.

Therefore, set $r = R_{\text{tube}} = R$ and $V = 0$ to solve for C_1. From Eq. (1.8):

$$0 = \frac{1}{2\mu}\frac{dP}{dx}\frac{R^2}{2} + C_1 \tag{1.9}$$

$$0 = \frac{1}{2\mu}\frac{dP}{dx}\frac{R^2}{2} + C_1 \quad \text{therefore} \quad C_1 = -\frac{1}{2\mu}\frac{dP}{dx}\frac{R^2}{2} \tag{1.10}$$

The equation, which gives velocity as a function of radius, r, is then:

$$V = \frac{1}{4\mu}\frac{dP}{dx}[r^2 - R^2] \tag{1.11}$$

The fifth and final assumption for this development is that the flow is laminar and not turbulent. Otherwise, this parabolic velocity profile would not be a good representation of the velocity profile across the cross-section.

Note that in Eq. (1.11), dP/dx must have a negative value for pressure drops that cause a positive velocity. This is consistent with the definition of positive x as a value to the right, or downstream. Note also that the maximum velocity will occur on the arterial centerline. By substituting $r = 0$ into Eq. (1.11) the maximum velocity is obtained and may be written as shown below in Eq. (1.12).

$$V_{\text{max}} = \frac{1}{4\mu}\frac{dP}{dx}[-R^2] \tag{1.12}$$

Furthermore, velocity as a function of radius may be written in the more convenient form:

$$V = V_{\text{max}}\left[1 - \frac{r^2}{R^2}\right] \tag{1.13}$$

1.4.3 Flow rate

Now that the flow profile is known and a function giving velocity as a function of radius can be written, it is possible to integrate velocity multiplied by differential area to find the total flow passing through the tube. Consider a ring of fluid at some distance r, from the centerline of the tube. See Fig. 1.13.

The differential flow, dQ, passing through this ring may be designated as:

$$dQ = 2\pi\, rdr \tag{1.14}$$

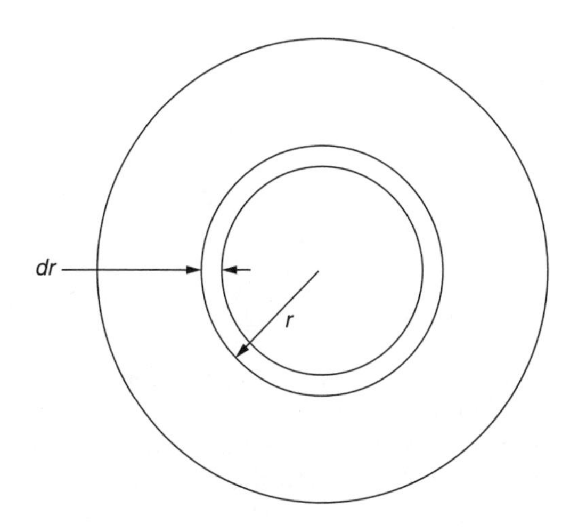

Figure 1.13 Shows a cross-section of a tube showing a ring of fluid at radius r. The thickness of the ring is *dr*.

To obtain the total flow passing through the cross-section, integrate the differential flow for $r = 0$ to $r = R_{\text{tube}}$ after substituting Eq. (1.11) for V.

$$Q = 2\pi \int_0^R Vrdr \tag{1.15}$$

$$Q = \frac{\pi}{2\mu}\frac{dP}{dx}\int_0^R (r^3 - rR^2)dr = \frac{\pi}{2\mu}\frac{dP}{dx}\int_0^R \left(\frac{r^4}{4} - \frac{r^2R^2}{2}\right)\Bigg|_0^R \tag{1.16}$$

This yields the expression:

$$Q = \frac{-\pi R^4}{8\mu}\frac{dP}{dx} \xrightarrow{\text{i.e}} \text{Poiseuille's law!} \tag{1.17}$$

Finally, it is now possible to solve for the average velocity across the cross-section and to check assumption 5, that the flow is laminar. The expression for the average velocity across the cross-section is

$$V_{\text{avg}} = \frac{Q}{A} = \frac{-R^2}{8\mu}\frac{dP}{dx} = \frac{V_{\text{max}}}{2} \tag{1.18}$$

To check assumption 5, use V_{avg} to calculate the Reynolds number and check to be sure that it is less than 2000. If the Reynolds number is greater than 2000 the flow may be turbulent and Poiseuille's law no longer applies.

Check: Is $\text{Re} = \dfrac{\rho VD}{\mu} < 2000$?

1.5 Bernoulli Equation

For inviscid, incompressible flow along a streamline, the Bernoulli equation can be used to investigate the relationship between pressures and velocities. This is particularly useful in situations with converging flows. The equation, named for the Swiss mathematician and physician, Daniel Bernoulli, is written:

$$P_1 + \frac{1}{2}\rho V_1^2 + \rho g z_1 = P_2 + \frac{1}{2}\rho V_2^2 + \rho g z_2$$

P_1 and P_2 are pressures at points 1 and 2. V_1 and V_2 are velocities at points 1 and 2, and z_1 and z_2 are heights at points 1 and 2.

Because the Bernoulli equation does not take frictional losses into account, it is not appropriate to apply the Bernoulli equation to flow through long, constant cross-section pipes as described by Poiseuille's law.

1.6 Conservation of Mass

If we want to measure mass flow rate in a very simple way, we might cut open an artery and catch blood in a beaker for a specific length of time. After weighing the blood in the beaker to determine the mass, m, we would divide by the time required to collect the blood, t, to get to flow rate, $Q = m/t$. However, this is not always a practical method of measuring flow rate.

Instead, consider that for any control volume, conservation of mass guarantees that the mass entering that control volume in a specific time is equal to the mass leaving that same control volume over the same time interval plus the increase in mass inside the control volume.

$$\left(\frac{\text{mass}}{\text{time}}\right)_{\text{in}} \Delta t = \left(\frac{\text{mass}}{\text{time}}\right)_{\text{out}} \Delta t + (\text{mass increase})$$

The continuity equation represents conservation of mass in a slightly different form, for cases where the mass inside the control volume does not increase or decrease and it is written:

$$\rho_1 A_1 V_1 = \rho_2 A_2 V_2 = \text{constant}$$

where ρ_1 is the fluid density at point 1, A_1 is the area across which fluid enters the control volume, and V_1 is the average velocity of the fluid across A_1. The variables ρ_2, A_2, and V_2 are density, area, and average velocity at point 2.

For incompressible flows under circumstances where mass inside the control volume does not increase or decrease, the density of the fluid is

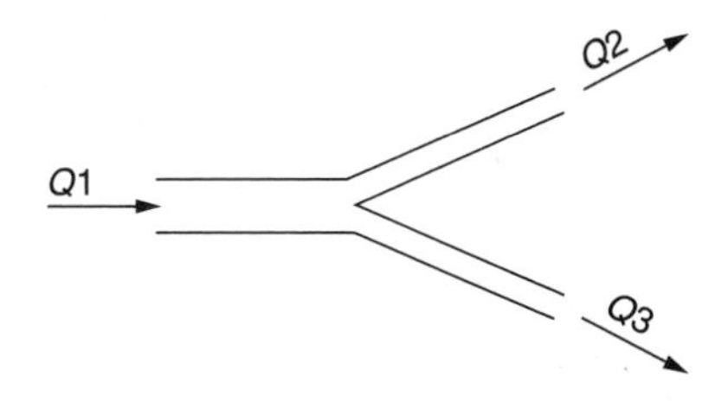

Figure 1.14 Shows a bifurcation in an artery.

constant ($\rho_1 = \rho_2$) and the continuity equation can be written in its more usual form:

$$A_1V_1 = A_2V_2 = Q = \text{constant}$$

where Q is the volume flow rate going into and out of the control volume.

Consider the flow at a bifurcation, or branching point, in an artery. See Fig. 1.14.

If we apply the continuity equation, the resulting expression is.

$$Q_1 = Q_2 + Q_3 \qquad \text{or} \qquad A_1V_1 = A_2V_2 + A_3V_3$$

1.6.1 Venturi meter example

A Venturi meter measures flow using the pressure drop between two points on either side of a Venturi throat as shown below in Fig. 1.15.

We can apply the Bernoulli equation between points 1 and 2, as well as, the continuity equation as follows. Assume that the Venturi meter is horizontal and that $z_1 = z_2 = 0$.

Bernoulli equation:

$$P_1 + \frac{1}{2}\rho V_1^2 + gz_1 = P_2 + \frac{1}{2}\rho V_2^2 + gz_2$$

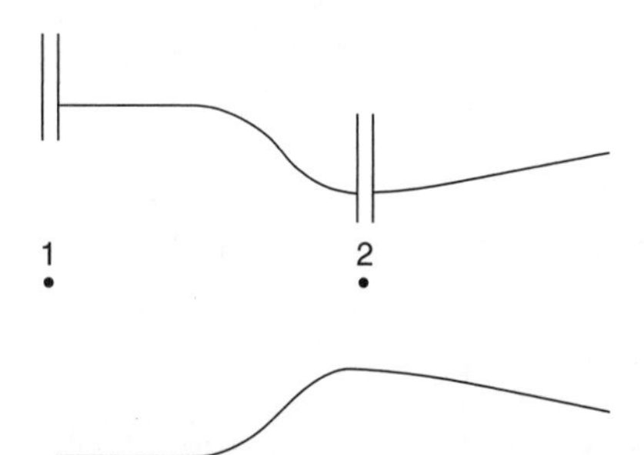

Figure 1.15 A Venturi flow meter with pressure measuring ports at points 1 and 2.

Continuity equation:

$$A_1V_1 = A_2V_2 = Q \quad \text{or} \quad V_2 = \frac{A_1}{A_2}V_1$$

Combining 1 and 2:

$$P_1 - P_2 = \frac{1}{2}\rho V_1^2\left[\left(\frac{A_1}{A_2}\right)^2 - 1\right]$$

Therefore, solving for V_1:

$$V_1 = \sqrt{\frac{2(P_1 - P_2)}{\rho\left[\left(\frac{A_1}{A_2}\right)^2 - 1\right]}}$$

Once we have solved for V_1, the ideal flow can be calculated by applying the continuity equation:

$$Q_{\text{ideal}} = V_1A_1$$

However, because of frictional losses through the Venturi, it will be necessary to calibrate the flow meter and adjust the actual flow using the calibration constant c:

$$Q_{\text{actual}} = cV_1A_1$$

1.7 Example Problem: Fluid Statics

In general, fluids exert both normal and shearing forces. This section reviews a class of problems in which the fluid is at rest. A velocity gradient is necessary for the development of a shearing force, so in the case where acceleration is equal to zero, only normal forces occur. These normal forces are also known as hydrostatic forces.

In Fig. 1.16 a point in a fluid P_1 is shown at a depth of h below the surface of the fluid. The pressure exerted at a point in the fluid by the column of fluid above the point is

$$P_1 = \gamma h$$

where P_1 is the pressure at point 1, γ is the specific weight of the fluid, and h is the distance between the fluid surface and the point P_1, or

$$P_1 = \rho g h$$

where ρ is fluid density, g is the acceleration due to gravity, and h is the distance between the fluid surface and the point P_1.

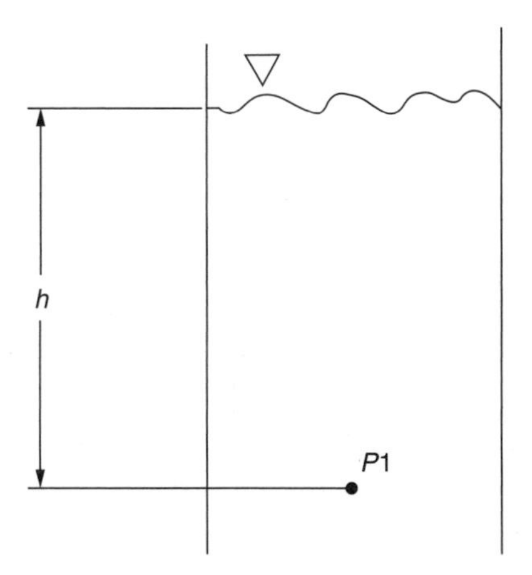

Figure 1.16 Fluid in a reservoir showing the depth of point P_1.

1.7.1 Example problem: fluid statics

A liquid (viscosity = 0.002 Ns/m^2; density = 1000 kg/m^3) is forced through the circular tube as shown in Fig. 1.17. A differential manometer is connected to the tube as shown to measure the pressure drop along the tube. When the differential reading h is 6 mm, what is the pressure difference between point 1 and point 2?

Solution. Assume that the pressure at point 1 is P_1. See Fig. 1.18. The pressure at point a, is increased by the column of water between point 1 and point a. The specific gravity of the water is 9810 N/m^3. If the vertical distance between 1 and a is d_a, then the pressure at a is

$$P_a = P_1 + g_{\text{water}}\, d_a$$

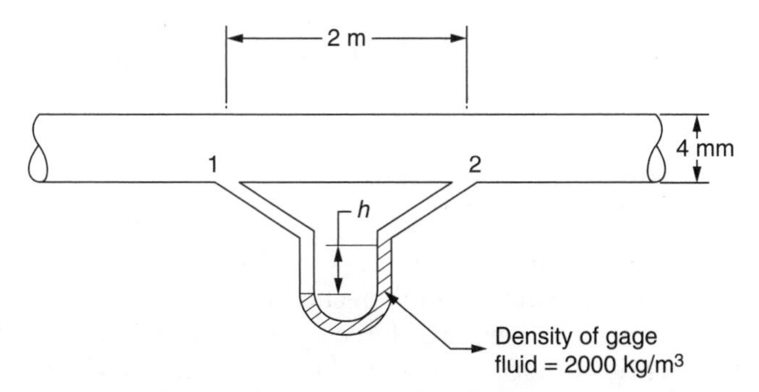

Figure 1.17 Pipeline for example problem 1.7.1.

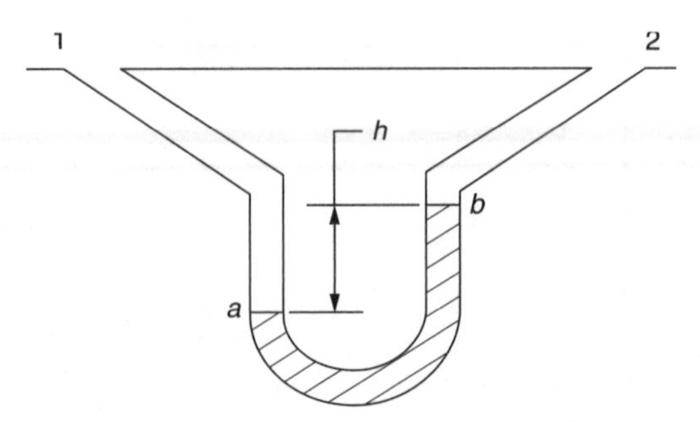

Figure 1.18 Enlargement of the manometer for example problem 1.7.1.

The pressure at *b* can now be calculated from the pressure at *a*. Point *b* is higher than point *a*, and therefore the pressure at b is less than the pressure at a. The pressure at *b* is:

$$\begin{aligned} P_b &= P_a - \gamma_{\text{gage}} \times h \\ &= P_1 + \gamma_{\text{water}}\, d_a - \gamma_{\text{gage}} \times h \end{aligned}$$

where g_{gage} is the specific weight of the gage fluid and *h* is the height of point *b* with respect to point *a*.

The pressure at 2 is also lower than the pressure at *b* and is calculated by:

$$\begin{aligned} P_2 &= P_b - (d_a - h)\gamma_{\text{water}} \\ &= P_1 + \gamma_{\text{water}}\, d_a - \gamma_{\text{gage}} \times h - (da - h)\gamma_{\text{water}} \end{aligned}$$

Finally, the pressure difference between 1 and 2 is written as

$$P_1 - P_2 = (\gamma_{\text{gage}} - \gamma_{\text{water}}) \times h$$

$$\Delta P = \left(2000(9.81)\frac{N}{m^3} - 1000(9.81)\frac{N}{m^3}\right)\frac{6}{1000}\,m = 58.9\frac{N}{m^2}$$

1.8 The Wormersley Number, α, a Frequency Parameter for Pulsatile Flow

The Wormersley number, or alpha parameter, is another dimensionless parameter that is used in the study of fluid mechanics. This parameter represents a ratio of transient to viscous forces, just as the Reynolds number represented a ratio of inertial to viscous forces. A characteristic frequency represents the time dependence of the parameter. The Wormersley number may be written by:

$$\alpha = \rho\sqrt{\frac{\omega}{\nu}} \quad \text{or} \quad \alpha = \rho\sqrt{\frac{\omega\rho}{\mu}}$$

where ρ = the vessel radius
ω = the fundamental frequency[3]
ρ = the density of the fluid
μ = the viscosity of the fluid
ν = the kinematic viscosity

In higher frequency flows, the flow profile is blunter near the centerline of the vessel since inertia becomes more important than viscous forces. Near the wall, where V is close to zero, viscous forces are still important.

Some typical values of the α parameter for various species are as follows:

- Human aorta, (#) $\alpha = 20$
- Canine aorta, (#) $\alpha = 14$
- Feline aorta, (#) $\alpha = 8$
- Rat aorta, (#) $\alpha = 3$

[3]The fundamental frequency is typically the heart rate. The units must be rad/s for dimensional consistency.

Chapter

2

Cardiovascular Structure and Function

2.1 Introduction

One may question why it is important for a biomedical engineer to study physiology. To answer this question, we could begin by recognizing that cardiovascular disorders are now the leading cause of death in developed nations. Furthermore, to understand the pathologies or dysfunctions of the cardiovascular system, engineers must first begin to understand the physiology or proper functioning of that system. If an engineer would then like to design devices and procedures to remedy those cardiovascular pathologies, then she/he must be well acquainted with physiology.

The cardiovascular system consists of a heart, arteries, veins, capillaries, and lymphatic vessels. Lymphatic vessels are vessels, which collect extracellular fluid and return it to the circulation. The most basic functions of the cardiovascular system are to deliver oxygen, to deliver nutrients, to remove waste, and to regulate temperature.

The heart is actually two pumps: the left heart and the right heart. The amount of blood coming from the heart (cardiac output) is dependent on arterial pressure. This chapter deals with the heart and its ability to generate arterial pressure in order to pump blood.

The adult human heart has a mass of approximate 300 g. If it beats 70 times per minute, then it will beat ~100,000 times per day, ~35 million times per year, and ~3 billion times (3×10^9) during your lifetime. If each beat ejects 70 mL of blood, your heart pumps over 7000 L, or the equivalent of 1800 gal per day. That is the same as 30 barrels of blood, each and every day of your life! The lifetime equivalent work done by the heart is the equivalent of lifting 30 tons to the top of Mount Everest. All of this work is done by a very hard working 300 g of muscle that do not rest!

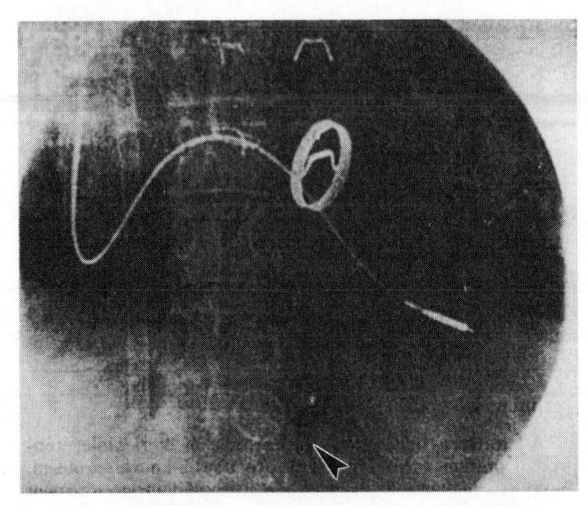

Figure 2.1 X-ray film showing escaped strut and disk (arrow). (Figure courtesy of Georg Thieme Verlag KG, *Thorac Cardiovasc Surg*. 1993;41:78)

2.2 Clinical Features

A 58-year-old German woman had undergone surgery in 1974 to fix a mitral stenosis.[1] She later developed a restenosis that caused the mitral valve to leak. She underwent surgery in November 1982 to replace her natural mitral valve with a Björk-Shiley convexo-concave mitral prosthesis.[2]

In July 1992, she suffered sudden and severe difficulty in breathing and was admitted to the hospital in her hometown in Germany. She had fluid in her lungs and the diagnosis was that she had a serious hemodynamic disturbance from inadequate heart function (cardiogenic shock). The physician listened for heart sounds but did not hear the normal click associated with a prosthetic valve. The patient's heart rate was 150 beats per minute, her systolic blood pressure was 50 mmHg, and her arterial oxygen saturation was 78 percent.

Echocardiography[3] was performed and no valve disk was seen inside the valve ring. An x-ray (see Fig. 2.1) appeared to show the valve disk in the abdominal aorta and the outflow strut in the pulmonary vein. Emergency surgery was performed, the disk and outflow strut were

---

[1] Narrowing or stricture.

[2] Artificial body part.

[3] Graphically recording the position and motion of the heart walls and internal structures.

removed and a new Medtronic-Hall mitral valve prosthesis was implanted into the woman and the patient eventually recovered.

2.3 Functional Anatomy

The cardiovascular system can be further divided into three subsystems. The systemic circulation, the pulmonary circulation, and the coronary circulation are the subsystems that, along with the heart and lungs, make up the cardiovascular system. See Fig. 2.2.

The three systems can be divided functionally by the tissue to which they supply oxygenated blood. The systemic circulation is the subsystem that is supplied by the aorta and that feeds the systemic capillaries. The pulmonary circulation is the subsystem supplied by the pulmonary artery that feeds the pulmonary capillaries. The coronary circulation is the specialized blood supply that perfuses cardiac muscle.

The cardiovascular system has four basic functions:

1. It supplies oxygen to body tissues.
2. It supplies nutrients to those same tissues.

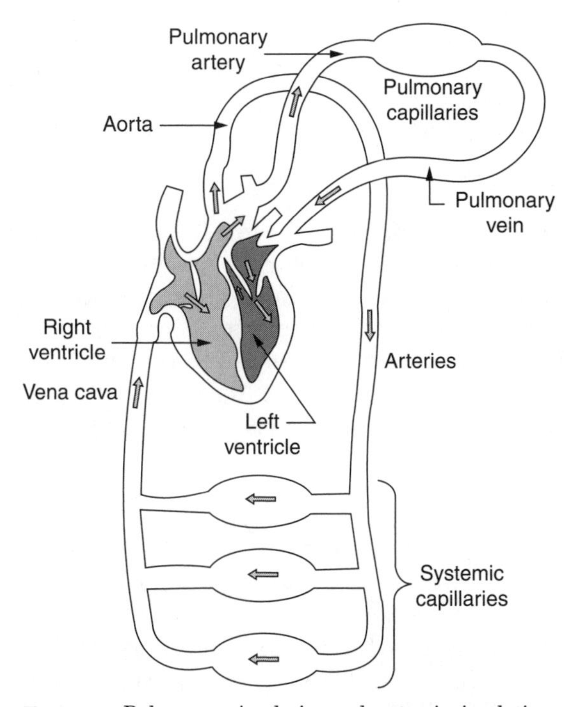

Figure 2.2 Pulmonary circulation and systemic circulation.

3. Removal of carbon dioxide and other wastes from the body.
4. The fourth function is that of temperature regulation.

The path of blood flowing through the circulatory system and the pressures of the blood at various points along the path tell us much about how tissue is perfused with oxygen. The left heart supplies oxygenated blood to the aorta at a relatively high pressure.

Blood continues along the path through the circulatory system and its path may be described as follows. Blood flows into smaller arteries and finally into systemic capillaries where oxygen is supplied to the surrounding tissues. At the same time, the blood picks up waste carbon dioxide from that same tissue and continues flowing into the veins. Eventually, the blood returns to the vena cava. From the vena cava, deoxygenated blood flows into the right heart. From the right heart, blood that is still deoxygenated flows into the pulmonary artery. The pulmonary artery supplies blood to the lungs where carbon dioxide is exchanged for oxygen. Blood, which has been resupplied with oxygen, flows from the lungs, through the pulmonary veins and back to the left heart.

It is interesting to note that blood flowing through the pulmonary artery is deoxygenated blood and blood flowing through the pulmonary vein is oxygenated blood. Although systemic arteries carry oxygenated blood, it is a mistake to think of arteries only as vessels which carry oxygenated blood. A more appropriate distinction between arteries and veins is that arteries carry blood at a relatively higher pressure than within their corresponding veins.

2.4 The Heart as a Pump

The heart is a four-chambered pump and is the driving force for moving blood through the circulatory system. The four chambers can be broken down into two upper chambers or atria, and two lower chambers, known as ventricles. Check valves between the chambers ensure that the blood moves in only one direction and enables the pressure in the aorta, for example, to be much higher than the pressure in the lungs, without blood flowing backwards from the aorta toward the lungs.

The four chambers of the heart including the right atrium, right ventricle, left atrium, and left ventricle are shown in Fig. 2.3. Blood enters the right atrium from the vena cava. From the right atrium, blood is pumped into the right ventricle. From the right ventricle, blood is pumped downstream through the pulmonary artery, to the lungs where it is enriched with oxygen and gives up carbon dioxide.

On the left side of the heart, oxygen enriched blood enters the left atrium from the pulmonary vein. When the left atrium contracts it pumps blood into the left ventricle. When the left ventricle contracts it

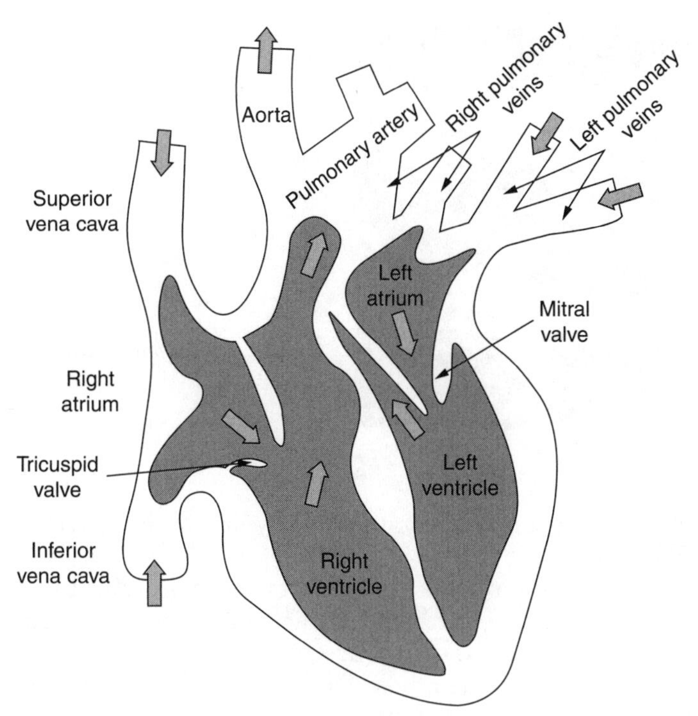

Figure 2.3 Heart chambers and flow through the heart. Arrows indicate direction of blood flow.

pumps blood to a relatively high pressure and ejects the blood from the left ventricle into the aorta.

2.5 Cardiac Muscle

The myocardium is composed of millions of elongated, striated, multinucleated cardiac muscle cells. These cells are approximately 15 mm × 15 mm × 150 mm long and can be depolarized and repolarized like skeletal muscle. Figure 2.4 below shows a typical group of myocardial muscle cells.

Individual cardiac muscle cells are interconnected by dense structures known as intercalated disks. The cells form a latticework of muscular tissue known as a syncytium.

A multinucleate mass of cardiac muscle cells form a functional syncytium (pronounced sin–sish′e-um). The heart has two separate muscle syncytia. The first is the mass that makes up the two atria and the second is the muscle mass that makes up the two ventricles. Fibrous rings that surround the valves between atria and ventricles separate the two syncytia. When one muscle mass is stimulated, the action potential spreads over the entire syncytium. As a result, under normal

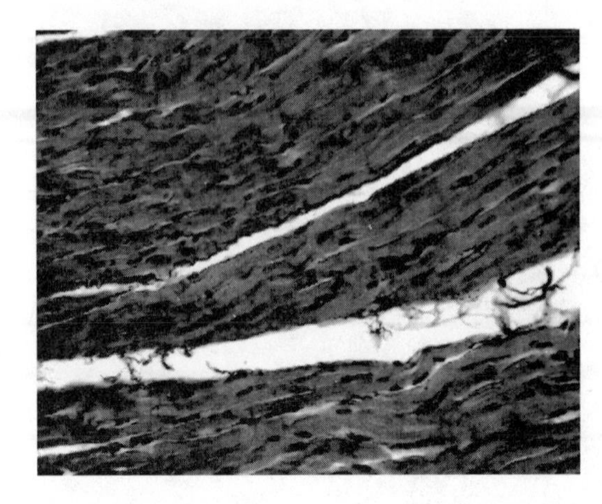

Figure 2.4 Photomicrograph of myocardial muscle.

circumstance, both atria contract simultaneously and both ventricles contract simultaneously.

Contraction in myocardium takes 10 to 15 times as long as it takes in average skeletal muscle. Myocardium contracts more slowly because sodium/calcium channels in myocardium are much slower than sodium channel in skeletal muscle during repolarization. In addition, immediately after the onset of the action potential, the permeability of cardiac muscle membrane for potassium ions decreases about five-fold. This effect does not happen in skeletal muscle. This decrease in permeability prevents a quick return of the action potential to its resting level.

2.5.1 Biopotential in myocardium

The cellular membranes of myocardial cells are polarized in the resting state like any other cells in the body. The resting, transmural electrical potential difference is approximately −90 mV in ventricular cells. The inside of the cell is negative with respect to the outside. This transmembrane potential exists because the cell membrane is selectively permeable to charged particles. Figure 2.5 shows the transmembrane resting potential in a cardiac cell.

The principal electrolyte ions inside of myocardial cells, which are responsible for the transmembrane potential, are sodium, potassium, and chloride. Negative anions associated with proteins and other large molecules are also very important for the membrane potential. They attract the positive potassium ion that can go inside the cell.

The permeability of the membrane to sodium ions is very low at rest, and the ions cannot easily pass through the membrane. On the other hand, permeability of the membrane with respect to both potassium and chloride ions is much higher and those ions can pass relatively easily

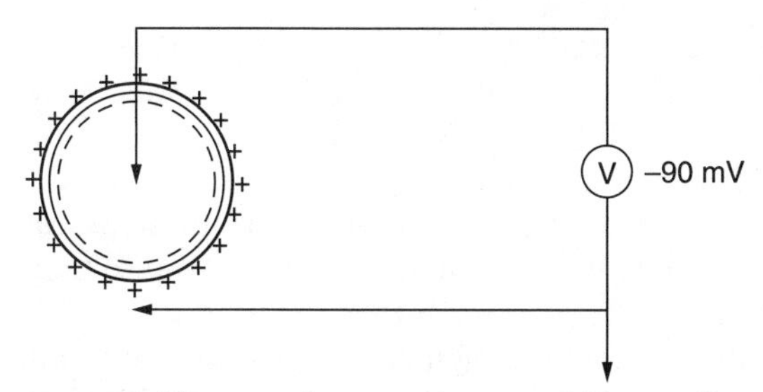

Figure 2.5 Transmembrane resting potential in a cardiac cell.

through the membrane. Since sodium ions cannot get inside the cell and the concentration of positively charged sodium ions is higher outside of the cell, a net negative electrical potential can be measured across the cell membrane.

The net negative charge inside the cell also causes potassium to concentrate inside the cell to counterbalance the transmembrane potential difference. However, the osmotic pressure caused by the high concentration of potassium prevents a total balancing of the electrical potential across the membrane.

Since some sodium continually leaks into the cell, maintaining steady state balance requires continual active transport. An active sodium-potassium pump in the cell membrane uses energy to pump sodium out of the cell and potassium into the cell.

2.5.2 Excitability

Excitability is the ability of a cell to respond to an external excitation. When the cell becomes excited the membrane permeability changes allowing sodium to freely flow into the cell. In order to attempt to maintain equilibrium, potassium that is at a higher concentration inside the cell, flows to the outside. In Fig. 2.6 a cardiac muscle cell is shown depolarizing.

In order to obtain a regulated depolarization, it is crucial that the increases in sodium permeability and in potassium permeability are offset in time. The sodium permeability must increase at the beginning

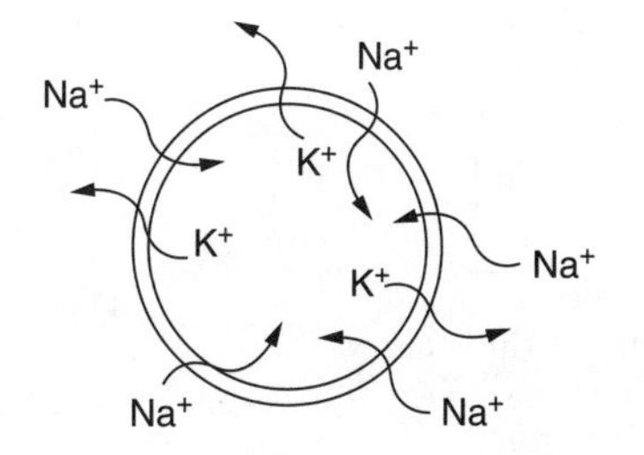

Figure 2.6 Depolarization of a myocardial cell.

of depolarization and the potassium channel then increases during repolarization. It is also important that the potassium channels that open during an action potential are different from the leak channels that allow potassium to pass through the membrane at rest. In cardiac muscle, the action potential is carried mainly by calcium from the extracellular space rather than by sodium. This calcium is then used to trigger the release of intracellular calcium to initiate contraction.

The ability of a cell to respond to excitation depends on the elapsed time since the last contraction of that cell. The heart is refractory to stimulation until it has recovered from the previous stimulation. That is, if you apply stimulation below threshold, before the refractory period has passed, the cell will not give a response. In Fig. 2.7, the effective refractory period (ERP) for a myocardial cell is shown to be approximately 200 ms. After the relative refractory period (RRP) the cell is able to respond to stimulation if the stimulation is large enough. The time for the relative refractory period in the myocardium is approximately another 50 ms.

During a short period following the refractory period, a period of supernormality (SNP) occurs. During the period of supernormality, the cell's transmembrane potential is slightly higher than its resting potential.

The refractory period in heart muscle is much longer than in skeletal muscle, because repolarization is much slower. Ventricular muscle in dogs has a refractory period of 250 to 300 ms at normal heart rates. The refractory period for mammalian skeletal muscle is 2 to 4 ms and 0.05 ms for mammalian nerve fiber.

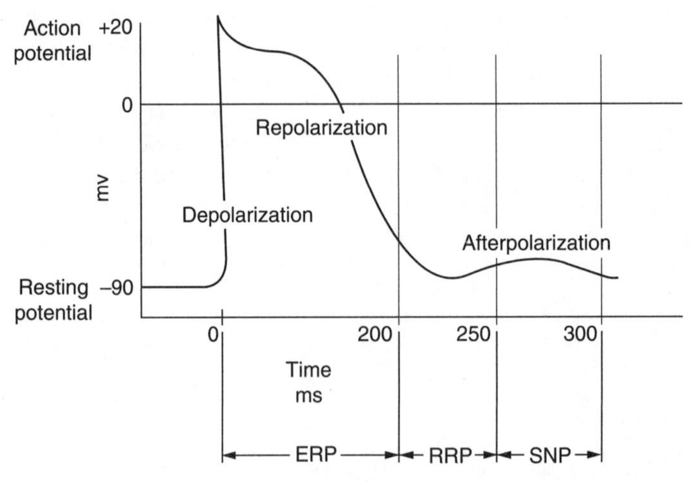

Figure 2.7 Action potential of a myocardial cell showing the effective refractory period (ERP), the relative refractory period (RRP), and the supernormality period (SNP).

2.5.3 Automaticity

Automaticity is the ability of a certain class of cells to depolarize spontaneously without external stimulation. All cardiac cells in the conduction system have automaticity. The sinoatrial (SA) node is a small group of cardiac muscle fibers on the posterior wall of the right atrium and the muscle cells that make up the SA node have especially strong automaticity. Under normal circumstances, the cells of the SA node control heart rate. They form the pacemaking site of the heart. The atrioventricular node (AV) is located at the lower interatrial septum. The AV node provides a limiting effect on the maximum heart rate. The AV node can only transmit impulses up to about 200 beats per minute. If the atrial rate is higher, as in atrial tachycardia, then some beats will be transmitted to the ventricle and others will not. This condition is known as atrioventricular block.

2.6 Heart Valves

Four cardiac valves help to direct flow through the heart. Heart valves cause blood to flow only in the desired direction. If a heart without heart valves were to contract, it would compress the blood causing it to flow both backward and forward (upstream and downstream). Instead, under normal physiological conditions, heart valves act as check valves to prevent blood from flowing in the reverse direction. In addition, heart valves remain closed until the pressure behind the valve is large enough to cause blood to move forward.

Each human heart has two atrioventricular valves that are located between the atria and the ventricles. The tricuspid valve is the valve between the right atrium and the right ventricle. The mitral valve is the valve between the left atrium and the left ventricle. The mitral valve prevents blood from flowing backwards into the pulmonary veins and therefore into the lungs, even when the pressure in the left ventricle is very high. The mitral valve is a bicuspid valve having two cusps and the tricuspid valve has three cusps.

The other two valves in the human heart are known as semilunar valves. The two semilunar valves are the aortic valve and the pulmonic valve. The aortic valve is located between the aorta and the left ventricle and when it closes, it prevents blood from flowing backward from the aorta into the left ventricle. An aortic valve is shown in Fig. 2.8. The pulmonic valve is located between the right ventricle and the pulmonary artery and when it closes it prevents blood from flowing backwards, from the pulmonary artery into the right ventricle.

Papillary muscles are cone shaped projections of myocardial tissue that connect from the ventricle wall to the chordae tendineae, which are tendons connected to the edge of atrioventricular valves. The chordae

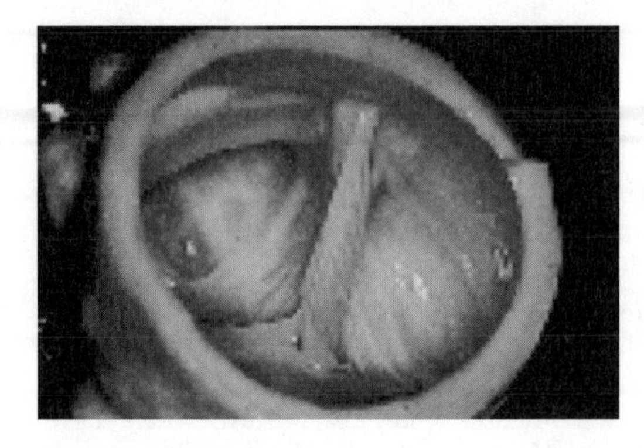

Figure 2.8 Aortic valve. (Reprinted with permission from Lingappa VR, Farey K, *Physiological Medicine*. New York: McGraw-Hill; 2000.)

tendineae are fine fibrous cords of collagen. These structures allow the AV valves to open and close, but they constrain the valves and prevent them from prolapsing, or collapsing backwards into the atria. Functionally, the papillary muscles and chordae tendineae are part of the valve complex with which they are associated. When the papillary muscles contract, they help maintain systolic close of the mitral or tricuspid valve.

2.6.1 Clinical features

Chordae tendineae rupture and papillary muscle paralysis can be consequences of a heart attack. This can lead to bulging of the valve, excessive backward leakage into the atria (regurgitation) and even valve prolapse. Valve prolapse is the condition under which the valve inverts backwards into the atrium. Because of these valve problems, the ventricle does not fill efficiently. Significant further damage, and even death, can occur within the first 24 h after a heart attack because of this problem.

2.7 Cardiac Cycle

Figure 2.9 is a graphical representation of the cardiac cycle during one heartbeat. Figure 2.10 also shows typical pressure in the aorta, left ventricle, and left atrium as a function of time.

Systole is the term used to describe the portion of the heartbeat during which ejection takes place. Figure 2.9 shows atrial systole occurring over a period of 0.1 s, just before ventricular systole. Ventricular systole is shown to occur over a time-period of 0.3 s, directly after atrial systole.

During atrial systole, pressure is generated in the atria, which forces the AV valves to open. The AV valves include the mitral valve and the tricuspid valves. The AV valves remain open during atrial systole, until ventricular systole generates enough ventricular pressure to force those valves to close.

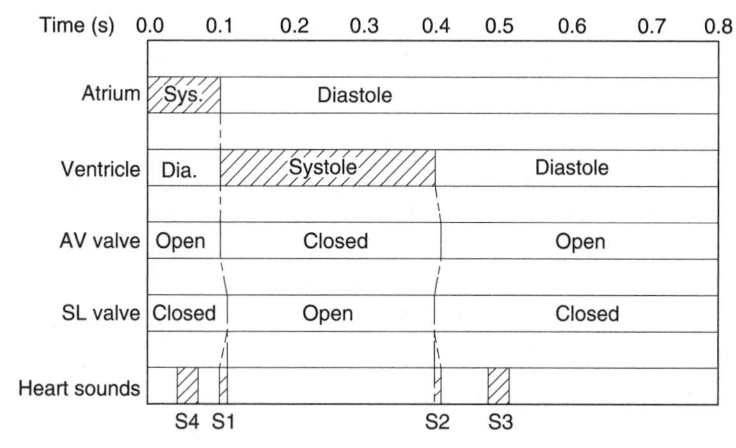

Figure 2.9 The timing of the cardiac cycle.

Diastole is the term used to describe the portion of the heartbeat during which chamber refilling is taking place. The ventricles undergo refilling, or ventricular diastole, during atrial systole. When the ventricle is filled and ventricular systole begins, then the AV valves are closed and the atria undergo diastole. That is, the atria begin refilling with blood.

Figure 2.9 shows a period of 0.4 s after ventricular systole, when both the atria and the ventricles begin refilling and both heart chambers are in diastole. During the period when both heart chambers are refilling, or undergoing diastole, both AV valves are open and both the aortic and pulmonic valves are closed. The aortic and pulmonic valves are also called semilunar valves.

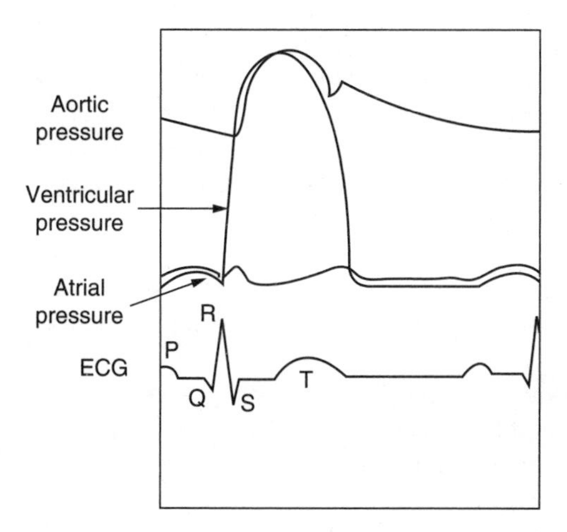

Figure 2.10 Typical pressures in the aorta, the left ventricle, and the left atrium are plotted versus time.

The semilunar valves are always closed except for a short period of ventricular systole when the pressure in the ventricle rises above the pressure in the aorta for the left ventricle and above the pressure in the pulmonary artery for the right ventricle. When the pressure in the left ventricle drops below the pressure level in the aorta, the pressure in the aorta forces the aortic valve to close, thus preventing blood from flowing backward into the left ventricle. Similarly, when the pressure in the right ventricle drops below the pressure levels in the pulmonary artery, the pressure in the pulmonary artery forces the pulmonic valve to close, preventing blood from flowing backward into the right ventricle.

2.8 Heart Sounds

Figure 2.9 also shows heart sounds associated with events during the cardiac cycle. Heart sound one is the "lub" associated with the closing of the AV valves. Heart sound two is the "dub" associated with the closing of the semilunar valves. Heart sounds are relatively low pitched. Most are between 10 and 500 Hz and of a relatively low intensity. Blood movement between atria and ventricles produces some components of sounds one and two, by vibration of the ventricular walls, and even by vibration of the base of the aorta and pulmonary arteries. A third, less distinct heart sound may occur shortly after the opening of the AV valves. This sound can be heard most often in children or in adults with thin chest walls. A fourth sound may occur during atrial systole, when a small volume of blood is injected into the nearly filled ventricle. Once again, this less distinct sound is heard most often in children or in thin adults.

2.8.1 Clinical features

An increased end-systolic volume is typical in heart failure. In a patient with a failing heart, rapid early ventricular filling produces an audible third heart sound termed S_3, which occurs after S_2.

Late ventricular filling with atrial contraction is normally silent. In patients with hearts that have become stiffer, the end-diastolic blood flow is against a noncompliant ventricle resulting in an audible fourth heart sound termed S_4. Heart sound S_4 would occur just prior to S_1.

The ECG is also plotted for timescale comparison. During ventricular systole, the ventricular pressure waveform rises rapidly and when the ventricular pressure rises above the aortic pressure, the aortic valve opens. The aortic valve remains open during the time when the aortic pressure is lower than the left ventricular pressure. When the ventricular pressure drops below the aortic pressure, the aortic valve closes to prevent back flow into the ventricle from the aorta.

Figure 2.11 shows a more detailed aortic pressure-time curve. Systolic pressure is the largest pressure achieved during systole. Diastolic pressure

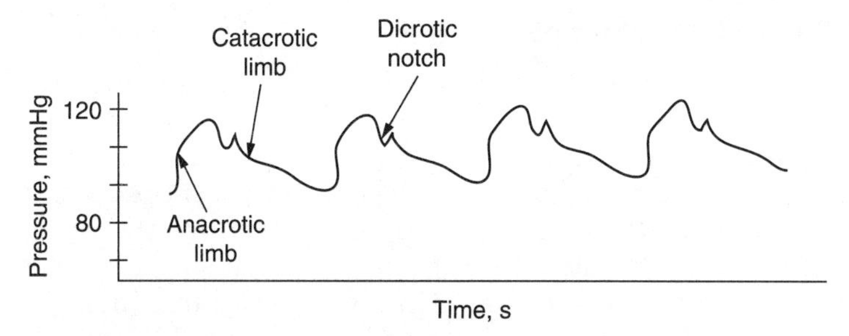

Figure 2.11 Aortic pressure waveform.

is the lowest aortic pressure measured during diastole. The pulse pressure is defined as the difference between the systolic and diastolic pressure. The anacrotic limb of the pressure waveform is the ascending limb of the pressure waveform and the descending portion of the waveform is known as the catacrotic limb. A notch exists in the normal catacrotic limb of the aortic pressure waveform. This notch is known as the dicrotic notch and is related to a transient drop in pressure due to the closing of the aortic valve.

Figure 2.12 shows a close-up view of a single pulse from an aortic pressure-time curve. The systolic pressure, diastolic pressure, and pulse pressure are shown graphically in the figure. It is possible to find the mean pressure over the beat by integrating the area under the curve and then dividing by the length of the time interval between t_1 and t_2. In the graph shown the mean pressure, P can be found by

$$P_{\text{mean}} = \frac{1}{t_2 - t_1} \int_{t_2}^{t_1} P \, dt$$

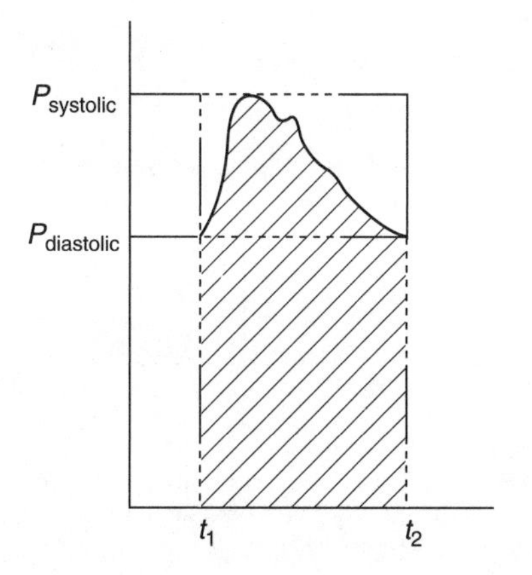

Figure 2.12 A single pressure waveform plotted against time.

2.9 Factors Influencing Flow and Pressure

The relationship between blood flow and blood pressure is one of the chief topics of this book. An engineer who is familiar with constant stroke volume pumps might imagine that flow from the heart is a simple function of heart rate. An engineer who designs pressure controllers may think of flow to a given capillary bed as related directly to the hydraulic resistance in that bed. In fact, the heart can sometimes be modeled as a constant stroke volume, variable-speed pump, but the system containing the heart and circulatory system is a complex network with a control system that includes pressure inputs, flow inputs, as well as neural and chemical feedback.

The pulse pressure in the aorta can be thought of chiefly as a function of heart rate, peripheral resistance, and stroke volume.

Figure 2.13 shows the effect of heart rate on pressure. If peripheral resistance and stroke volume remain constant, an increase in heart rate causes the heart to pump more blood into the aorta over a fixed time-period, and systolic blood pressure increases. At the same time, the increased rate of pumping leaves less time between heartbeats for the pressure in arteries to decrease, simultaneously the diastolic pressure also increases. As the heart rate increases, with fixed peripheral resistance and fixed stroke volume, cardiac output, or flow into the systemic circulation, increases.

On the other hand, if heart rate and peripheral resistance are constant, but stroke volume changes, pressure and flow are also affected. Figure 2.14 shows the effect of stroke volume on pressure. An increase in stroke volume increases pressure because of the extra volume of blood that is pumped into the aorta during each heart beat. However, a healthy human would not normally experience a case of constant heart rate and constant peripheral resistance with increasing stroke volume. As shown in the figure, under resting conditions heart rate drops and

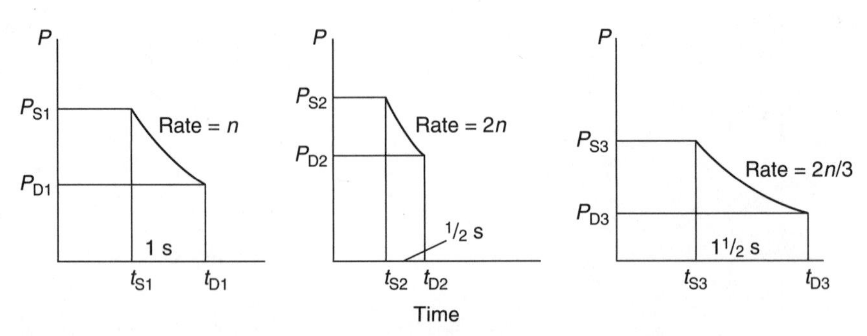

Figure 2.13 Aortic pressures as a function of heart rate. P_S is the systolic pressure and P_D is the diastolic pressure. Picture 2, the middle picture, shows the fastest heart rate and the picture on the right, picture 3, shows the slowest heart rate.

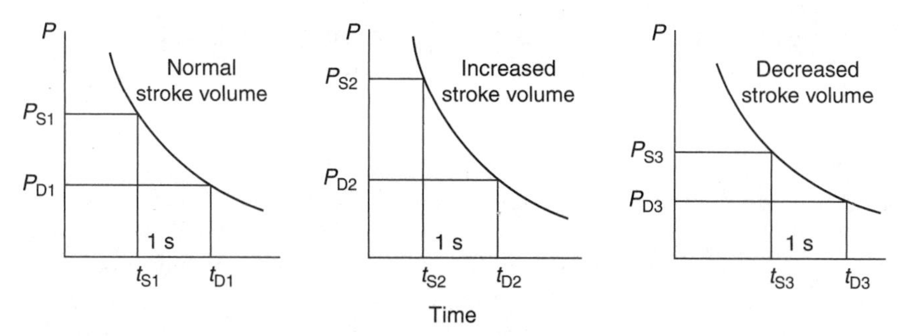

Figure 2.14 Shows the effect of stroke volume on pressure. P_S represents pressure at systole and P_D represents pressure at diastole.

peripheral resistance increases. The result is that a large stroke volume, as might be seen in well trained athletes, increases systolic pressure or perhaps maintains it at a steady level, while the decreased heart rate allows more time between beats and the diastolic pressure decreases. A very small stroke volume, as might be seen in a patient with severe heart damage, would normally result in an increased heart rate as the cardiovascular system attempts to maintain cardiac output although a very small amount of blood is ejected from the left ventricle during each heart beat.

The third parameter that controls blood pressure and flow is peripheral resistance. This peripheral resistance comes chiefly from the resistance found in capillary beds and the more capillaries that are open, the lower the resistance will be. If heart rate and stroke volume remain constant as the need for oxygenated blood increases in peripheral tissues, more capillaries open to allow more blood to flow into the tissue and this corresponds to a decrease in peripheral resistance. Figure 2.15 shows

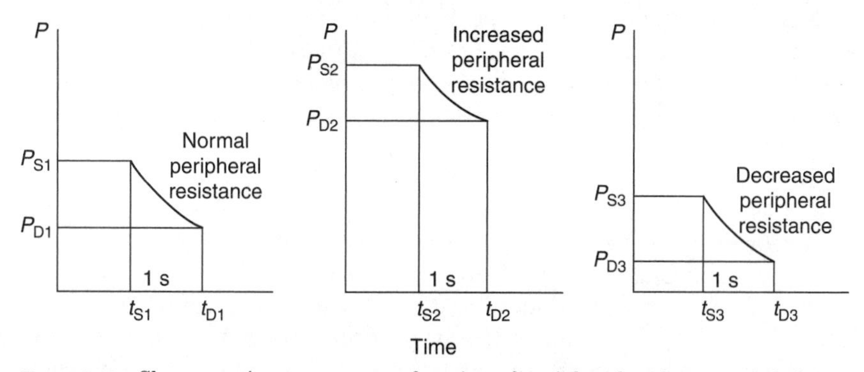

Figure 2.15 Shows aortic pressure as a function of peripheral resistance while heart rate and stroke volume are held constant.

that blood pressure increases with increasing peripheral resistance when heart rate and stroke volume are held constant.

Heart rate, stroke volume, and peripheral resistance are the primary factors that control the relationship between blood pressure and flow, or cardiac output, but there are other factors which contribute. For example, in the case of diseased heart valves, blood leakage backward through the valve into the heart can alter the relationship. This type of leakage through a valve is known as regurgitance. Regurgitance in heart valves has a similar effect to that of decrease in stroke volume. Blood that has been ejected into the aorta can flow backwards into the left ventricle through a faulty aortic valve resulting in a decreased effective stroke volume.

Arterial compliance, a characteristic that is related to vessel stiffness, can also affect the pressure flow relationship. Very stiff arteries increase the overall hydrodynamic resistance of the system. This increase in resistance causes a decrease in blood flow to the peripheral tissue. The complex control system in the cardiovascular system will result in compensation that increases heart rate and therefore increases blood pressure to compensate for the increased resistance.

2.10 Coronary Circulation

The blood supply to the heart is known as the coronary circulation. Because heart muscle requires a blood supply in order to be able to provide blood supply to the rest of the body, this system, which is relatively small by blood volume, provides a very important function. Approximately one-third of people in the countries of the western world die as a result of coronary artery disease. Almost all elderly people have some impairment of the coronary circulation.

The resting coronary blood flow in humans is around 225 mL/min, which is a bit less than 5 percent of cardiac output. During strenuous exercise, cardiac output in young healthy people increases four-to seven-fold. Because the heart is also providing this increased flow at a higher arterial pressure, the work of the heart is increased six to nine times over the resting work output. The coronary blood flow only increases six to nine fold to supply oxygen and nutrients needed by the cardiac muscle during this increased workload.

The main coronary artery arises from the root of the aorta. It then branches into the left and right coronary arteries. Seventy-five percent of the coronary blood supply goes through the left coronary artery and 25 percent through the right coronary artery.

The left coronary artery supplies blood to the left lateral portion of the left ventricle and the anterior (front) portion of the ventricular septum. The ventricular septum is the wall between the left and right ventricles. The left coronary artery then branches into the anterior descending coronary artery and the circumflex coronary artery. Those two branches provide blood supply to the left ventricle.

The right coronary artery supplies the right ventricle, and the posterior (back) portion of the ventricular septum. The right coronary artery also supplies blood to the specialized muscle fibers of the sinoatrial (SA) node that is located at the top-rear-right of the right atrium and is the primary electrical pacemaking sight of the heart.

Blood flows through the capillaries of the heart and returns through the cardiac veins to the coronary sinus. The coronary sinus is the main drainage channel of venous blood from the heart muscle. It is a groove on the back surface of the heart between the left atrium and ventricle and it drains into the right atrium. Most of the coronary blood flow from the left ventricle returns to the coronary circulation through the coronary sinus. Most coronary blood flow to the right ventricle returns through the small anterior cardiac veins that flow directly into the right atrium.

2.10.1 Control of the coronary circulation

Control of the coronary circulation is almost entirely through local arterial and arteriolar control. The coronary arteries and arterioles vasodilate in response to the cardiac muscle's need for oxygen. Whenever muscular activity of the heart increases, blood flow to the heart increases. Decreased activity also causes a decrease in coronary blood flow. As blood flows through the coronary arteries, about 70 percent of the oxygen in the arterial blood is removed as it passes through the heart muscle. This means that there is not much oxygen left in the coronary venous blood that could have been removed in activity that is more strenuous. Therefore, it is necessary that coronary flow increases as heart work increases.

2.10.2 Clinical features

When blockage of a coronary artery occurs gradually over months or years, collateral circulation can develop. This collateral circulation allows adequate blood supply to the heart muscle to develop through an alternative path. On the other hand, since atherosclerosis is typically a diffuse process, blockage of other vessels will also eventually occur. The end result is a devastating and sometimes lethal loss of heart function because the blood flow was dependent of a single collateral branch.

2.11 Microcirculation

Microcirculation is the flow of blood in the system of smaller vessels of the body whose diameters are 100 μm or less. This includes arterioles, metarterioles, capillaries, and venules. The microcirculation is involved chiefly in the exchange of gases, fluids, nutrients, and metabolic waste products.

2.11.1 Capillary structure

Capillaries are thin tubular structures whose walls are one cell layer thick. They consist of highly permeable endothelial cells. In the peripheral circulation there are about 10 billion capillaries with an average length of about 1 mm. The estimated surface area of all the capillaries in the entire system is about 500 m^2 and the total volume of blood contained in those capillaries is about 500 mL. Capillaries are so plentiful that it is rare that any cell in the body is more than 20 μm away from a capillary.

Consider the structure of blood vessels in terms of branching "generations." If the aorta branches or bifurcates into two, those branches may be considered the second generation. Arteries branch six or eight times before they become small enough to be arterioles. Arterioles then branch another two to five times, for a total of eight to thirteen generations of branches within arteries and arterioles. Finally, the arterioles branch into metarterioles and capillaries that are 5 to 9 micrometers (μm; also called microns) in diameter.

In the last arteriole generation, blood flows from the arterioles into the metarterioles, which are sometimes called terminal arterioles. From there the blood flows into the capillaries.

There are two types of capillaries. One type is the relatively larger preferential channel and the second type is the relatively smaller true capillary. At the location where each true capillary begins from a metarteriole, there is smooth muscle fiber surrounding the capillary known as the precapillary sphincter. The precapillary sphincter can open and close the entrance to a true capillary, completely shutting off blood flow through this capillary.

After leaving the capillaries most blood returns to venules and then eventually back into the veins. However, about 10 percent of the fluid leaving the capillaries enters the lymphatic capillaries and returns to the blood through the lymphatic system.

2.11.2 Capillary wall structure

The wall of a capillary is one cell layer thick and consists of endothelial cells surrounded on the outside by a basement membrane. The wall thickness of a capillary is about 0.5 μm. The inside diameter is approximately 4 to 9 μm, which is barely large enough for an erythrocyte, or red blood cell, to squeeze through. In fact, erythrocytes, which are approximately 8 μm in diameter, must often fold in order to squeeze through the capillary.

Capillary walls have intercellular clefts, which are thin slits between adjacent endothelial cells. The width of these slits is only about 6 to 7 nanometers (nm) (6 to 7×10^{-9} m) on the average. Water molecules and most water soluble ions diffuse easily through these pores.

The pores in capillaries are different sizes for different organs. For example, intercellular clefts in the brain are known as "tight junctions." These pores allow only extremely small molecules like water, oxygen, and carbon dioxide to pass through the capillary wall. In the liver, the opposite is true. The intercellular clefts in the liver are wide open and almost anything that is dissolved in the plasma goes through these pores.

Bibliography

Herse B, Baryalei M, Wiegand V, and Autschbach R. Fractured strut of Bjork-Shiley 70 degrees convexo-concave mitral valve prosthesis found in left coronary artery. *Thorac Cardiovasc Surg*. 1993;41:77–79.

Selkurt EE. *Basic Physiology for the Health Sciences*, 2nd ed. Boston, MA: Little, Brown & Co; 1982.

Guyton AC, Hall JE. *Textbook of Medical Physiology*, 10th ed. Philadelphia: W.B. Saunders; 2000.

Lingappa VR, Farey K. *Physiological Medicine*. New York: McGraw-Hill; 2000.

Milnor WR. *Cardiovascular Physiology*. Oxford: Oxford University Press; 1990.

Yoganathan A. Cardiac valve prostheses. In: Bronzino J, ed. *The Biomedical Engineering Handbook*. Boca Raton, FL: CRC Press; 1995:1847–1870.

Chapter

3

Pulmonary Anatomy, Pulmonary Physiology, and Respiration

3.1 Introduction

Three primary functions of the respiratory system include gas exchange, acid-base balance, and the production of sound. The most basic activity of the respiratory system is to supply oxygen for metabolic needs and to remove carbon dioxide. Carbon dioxide, which is carried in the hemoglobin of our red blood cells, is exchanged in the lungs for oxygen, which is also carried in the hemoglobin. Fresh air that we continuously inspire into the lungs is exchanged for air that has been enriched with carbon dioxide. An increase in carbon dioxide leads to an increase in hydrogen ion concentration resulting in lower pH or more acidity. The respiratory system aids in acid-base balance by removing CO_2 from the body. Phonation or the production of sounds by air movement through vocal cords is the third important function of the respiratory system. When you speak, the muscles of respiration cause sound to move over the vocal cords and through the mouth resulting in intelligible sounds.

To understand the path of air through the respiratory system, consider the system components, which include two lungs, a trachea, bronchi, bronchioles, alveoli, and their associated blood vessels. These components are shown in Fig. 3.1. Air enters the respiratory system through the mouth or nose. After being filtered and heated the air passes into the pharynx and the larynx and then the air enters the trachea where it passes to the bronchial tree. There are three generations of bronchi, which are conducting airways greater than about 0.5 cm in diameter. Bronchioles are also conducting airways that have diameters greater than approximately

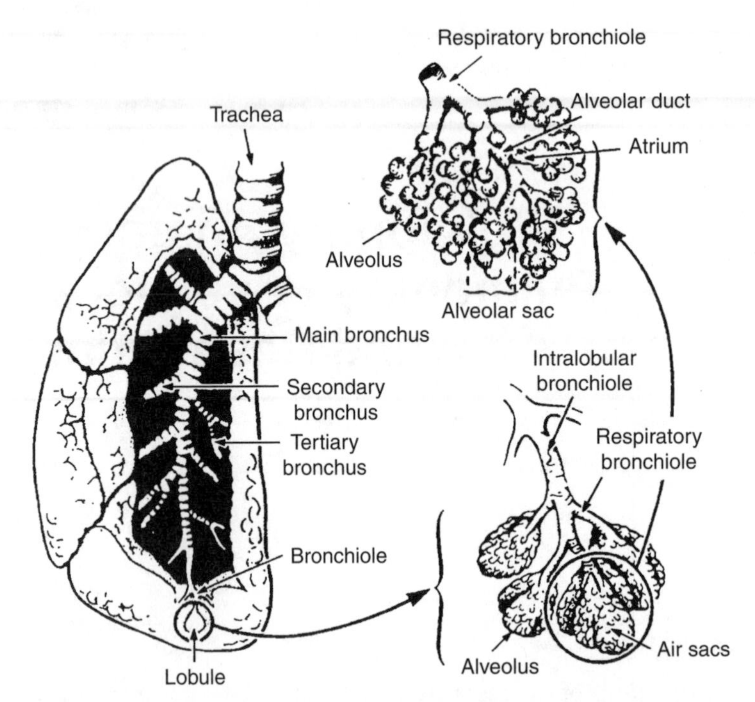

Figure 3.1 Functional anatomy of the lung. (Reprinted with permission from Selkurt E. *Basic Physiology for the Health Sciences*. Boston, Little, Brown; 1982.)

0.05 cm. There are approximately 13 generations of bronchioles with the last generation being known as terminal bronchioles.

Beyond the terminal bronchioles, air enters the transitional zone. There are several generations of respiratory bronchioles, just beyond the bronchi. Respiratory bronchioles have a diameter of approximately 0.05 cm and a length of about 0.1 cm. At the ends of the respiratory bronchioles, air passes into alveolar ducts and finally into alveolar sacs. Alveolar ducts and alveolar sacs make up the respiratory zone of the respiratory system. Figure 3.1 shows the functional anatomy of the lungs. At the left, you can see the gross structure of the right lung and the conducting airways. At the upper right, the branching of a bronchiole can be seen. Finally, at the lower right, at highest magnification, respiratory bronchioles, alveolar ducts, alveolar sacs, and alveoli are shown.

3.2 Clinical Features

Oxygen and carbon dioxide in blood are carried in hemoglobin, a substance inside the red blood cells. Hemoglobin in oxygenated blood that is returning from areas of the lung where gas exchange has occurred is normally saturated. Therefore hyperventilation of room air will not add much oxygen to the blood stream. If a patient suffers from low oxygen due to a mismatch in ventilation and perfusion, the patient must breathe air with a higher percentage of oxygen than that which occurs in room air.

3.3 Alveolar Ventilation

Alveolar ventilation is the exchange of gas between the alveoli and the external environment. It can be measured as the volume of fresh air entering (and leaving) the alveoli each minute. Oxygen from the atmosphere enters the lungs through this ventilation and carbon dioxide from the venous blood returns to the atmosphere. Physicians and biomedical engineers often discuss alveolar ventilation in terms of standard lung volumes. These standard lung volumes are represented graphically in Fig. 3.2.

3.3.1 Tidal volume

Tidal volume (TV) is the volume of ambient air entering (and leaving) the mouth and nose per minute during normal, unforced breathing. Normal tidal volume for a healthy, 70-kg adult is approximately 500 mL per breath, but this value can vary considerably during exercise.

3.3.2 Residual volume

Residual volume (RV) is the volume of air left in the lungs after maximal forced expiration. This volume is determined by a balance between muscle forces and the elastic recoil of the lungs. Residual volume is

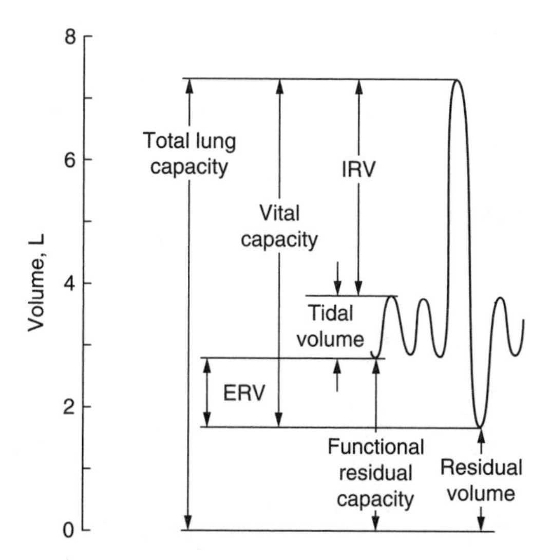

Figure 3.2 Standard lung volumes as measured by a spirometer.

approximately 1.5 L in a healthy, 70-kg individual. Residual volume is much greater for individuals with emphysema.

3.3.3 Expiratory reserve volume

Expiratory reserve volume (ERV) is the volume of air expelled from the lungs during a maximal forced expiration that begins at the end of normal tidal expiration. The expiratory reserve volume plus the reserve volume combine to make up the functional reserve capacity. Normal ERV for our standard 70-kg individual is approximately 1.5 L.

3.3.4 Inspiratory reserve volume

Inspiratory reserve volume (IRV) is the volume of air that is inhaled during forced maximal inspiration beginning at the end of normal tidal inspiration. This volume depends on muscle forces and also on the elastic recoil of the chest wall and the elastic recoil of the lungs. The IRV of a healthy 70-kg individual is approximately 2.5 L.

3.3.5 Functional residual capacity

Functional residual capacity (FRC) is the volume in the lungs at the end of normal tidal expiration. FRC depends on the equilibrium point at which the elastic inward recoil of the lungs balances the elastic outward recoil of the chest wall. Functional residual capacity consists of the sum of the residual capacity and the expiratory reserve volume and can be represented by Eq. (3.1). The FRC of a healthy 70-kg adult is approximately 3 L.

$$\mathrm{FRC} = \mathrm{RV} + \mathrm{ERV} \tag{3.1}$$

3.3.6 Inspiratory capacity

Inspiratory capacity (IC) is the volume of air taken into the lungs during a maximal inspiratory effort that starts at the end of a normal tidal volume expiration. Inspiratory capacity is the sum of TV and IRV and can be represented by Eq. (3.2). The IC of our healthy, 70-kg, adult is about 3 L.

$$\mathrm{IC} = \mathrm{IRV} + \mathrm{TV} \tag{3.2}$$

3.3.7 Total lung capacity

Total lung capacity (TLC) is the volume of air in the lungs after a maximal expiratory effort. The strength of contraction of inspiratory muscles, the inward elastic recoil of the lungs, and the chest wall determine the total lung capacity. The total lung capacity is also equal to the sum of the inspiratory reserve capacity, the tidal volume, the expiratory reserve capacity, and the reserve volume. A typical value for total lung capacity of a healthy 70-kg male is about 6 L. An equation representing the total lung capacity may be written by:

$$\mathrm{TLC} = \mathrm{IRV} + \mathrm{TV} + \mathrm{ERV} + \mathrm{RV} \tag{3.3}$$

3.3.8 Vital capacity

Vital capacity (VC) is the volume of air expelled from the lung during a maximal expiratory effort after a maximum forced expiration. The vital capacity is the difference between the total lung capacity and the reserve volume. Vital capacity in a normal healthy 70-kg adult is about 4.5 L. Equation (3.4) is for vital capacity:

$$\mathrm{VC} = \mathrm{TV} + \mathrm{IRV} + \mathrm{ERV} \tag{3.4}$$

3.4 Ventilation—Perfusion Relationships

Ventilation is the act of supplying air into the lungs and perfusion is the pumping of blood into the lungs. In this book we will use $\dot{V}$ to represent the air flow rate associated with ventilation and $\dot{Q}$ to represent the blood flow rate associated with perfusion. The ratio of ventilation to perfusion is important for lung function and is represented as the ventilation/perfusion ratio as shown in Eq. (3.5).

$$\text{Ventilation perfusion ratio} = \frac{\dot{V}}{\dot{Q}} \tag{3.5}$$

Pulmonary arterial smooth muscle vasoconstricts the vessels of the pulmonary capillary beds in response to hypoxia, or low oxygen. This type of vasoconstriction is one of the most important parameters that determine pulmonary blood flow. In other capillary beds within the body, smooth muscle vasodilates in response to tissue hypoxia, improving perfusion.

Resting ventilation is about 4 to 6 L/min. Resting pulmonary artery blood flow is about 5 L/min. At rest, therefore, the ventilation/perfusion ratio is about 0.8 to 1.2. Figure 3.3 shows a schematic representing a ventilation/perfusion ratio of 1 with a ventilation rate and perfusion rate both equal to 5 L/min.

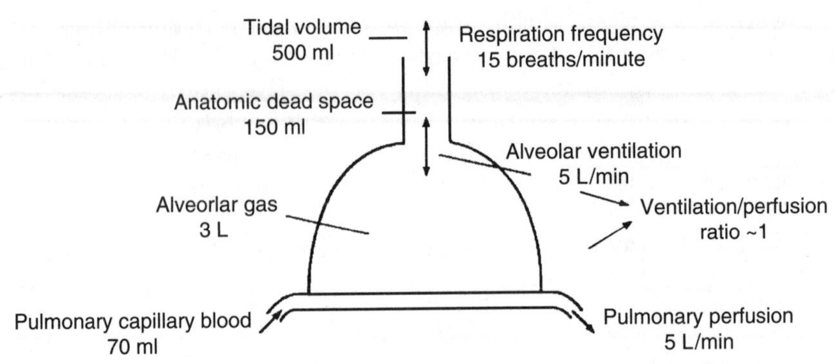

Figure 3.3 Pulmonary volumes and flows showing a ventilation perfusion ratio of 1.

3.5 Mechanics of Breathing

For the normal physiological case of breathing, air flows into the lungs when the alveolar pressure drops below the pressure of the surrounding ambient air. This is known as negative pressure breathing because the pressure in the alveoli must be negative with respect to the surrounding air. This negative pressure in the alveoli is caused by muscle contractions that increase the volume of the lung causing the alveoli to expand.

Transmural pressure gradient is defined by the difference in pressure between atmospheric air and the pressure in the alveoli. As the transmural pressure gradient increases, the alveoli expand.

Intrapleural pressure, which is also known as intrathoracic pressure, is caused by the mechanical interaction between the lung and chest wall. When all muscles of respiration are relaxed, left to themselves the lungs have a tendency to collapse whereas the chest wall tends to expand. This causes the intrapleural pressure to drop and this resulting negative pressure has the effect of holding the lung and the chest wall in close contact.

The primary muscles of breathing are the diaphragm, the external intercostals, and the accessory muscles. The diaphragm is a dome-shaped sheet of muscle with an area of about 250 cm^2. The diaphragm separates the abdominal cavity from the thoracic cavity.

During eupnea, or normal quiet breathing, in the supine position the diaphragm is responsible for two-thirds of the air entering the lungs.

3.5.1 Muscles of inspiration

The rib muscles (external intercostals) raise and enlarge the rib cage when contracted. The diaphragm and the rib muscles contract simultaneously during inspiration. If they did not, the contraction of the external intercostals could cause the diaphragm to be pulled upwards.

Accessory muscles are not used in normal quiet breathing. They are used however; during heavy breathing, as in exercise, and during the inspiratory phase of sneezing and coughing. An example of an accessory muscle is the sternocleidomastoid, which elevates the sternum to increase the anteroposterior (front to back) and the transverse (side to side) dimensions of the chest.

3.5.2 Muscles of expiration

Expiration is passive during quiet breathing and no muscle contraction is required. The elastic recoil of the alveoli due to alveolar stiffness is enough to raise the alveolar pressure above atmospheric pressure, the condition required for expiration.

Active expiration occurs during exercise, speech, singing, and the expiratory phases of coughing and sneezing. Active expiration may also be required due to pathologies such as emphysema. Muscles of active expiration are the muscles of the abdominal wall including the rectus abdominis, external and internal obliques, transverse abdominis, and internal intercostals muscles.

3.5.3 Compliance of the lung and chest wall

The slope of the pressure-volume curve of the lung is known as lung compliance. This volume compliance can be written as dV/dP, representing a change in volume per change in pressure. Compliance has the units of m^3/Pa and is inversely related to elasticity, or lung stiffness.

The pressure-volume curve for the lung is different for inspiration than it is for expiration. The difference in volume for a given pressure upon inspiration versus the same pressure upon expiration is known as hysteresis. At low lung volumes, the lung is stiffer (has a lower compliance) during inspiration than during expiration. At high lung volumes, the lung is less stiff (has a higher compliance) during inspiration.

3.6 Work of Breathing

The rate and depth at which one breathes, under normal circumstances is managed to minimize the amount of work that is done. If you try to breathe rapidly and shallowly for an extended period of time, you can transfer the necessary oxygen, but will rapidly grow tired from the effort. Work for a system that executes a cyclic process with only expansion and compression can be modeled by Eq. (3.6):

$$W_{a \to b} = \int_a^b P dV \tag{3.6}$$

In Eq. (3.6), W represents the work done between points a and b. P represents the pressure inside the system (in this case the lung). The volume of the system is represented by V. Figure 3.4 shows a typical pressure-volume curve during breathing. The work term can be thought of as an area under the pressure-volume curve.

For example, one part of the work done by the diaphragm on the lungs due to inspiration can be thought of as the work needed to overcome the elastic resisting forces of the chest wall and diaphragm. This work done to overcome elasticity can be represented by the area under the line AB as shown in Fig. 3.5.

The total work done by the diaphragm on the lungs due to inspiration can be thought of as the work needed to overcome the elastic resisting forces of the chest wall and diaphragm, plus the work done to overcome the resistance to flow. This total work of inspiration can be represented by the area under the inspiration curve as shown in Fig. 3.6.

During expiration, the elasticity of the lungs and chest wall help provide the stored energy so that diaphragm work is not necessary. The energy stored in these elastic tissues can be partially recovered during expiration. The work done to overcome resistance to flow cannot be overcome. The total work of one total breathing cycle,

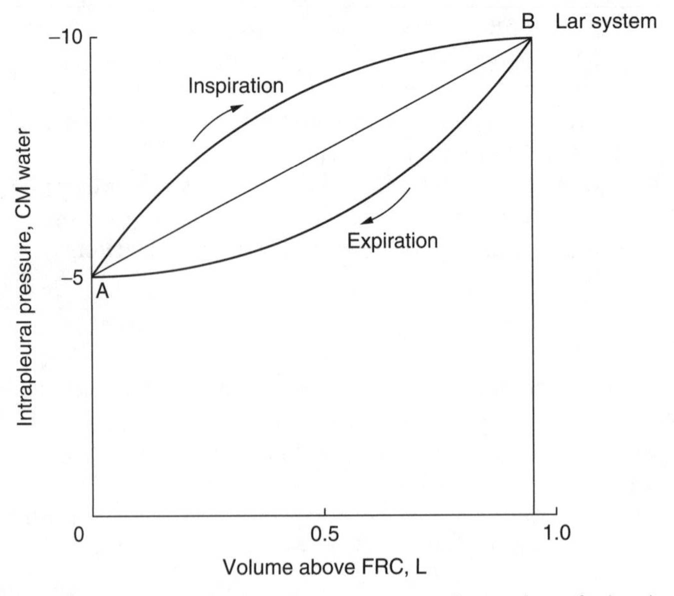

Figure 3.4 Shows intrapleural pressure versus lung volume for inspiration and expiration.

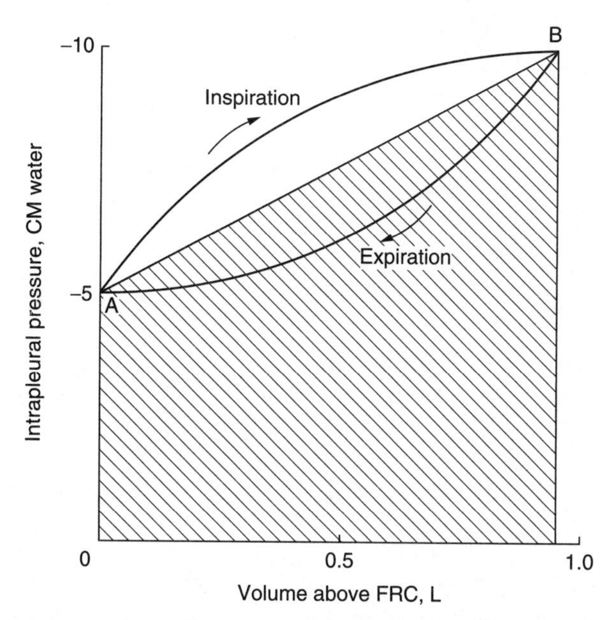

Figure 3.5 Pressure-volume curves showing the work done to overcome elasticity during inspiration.

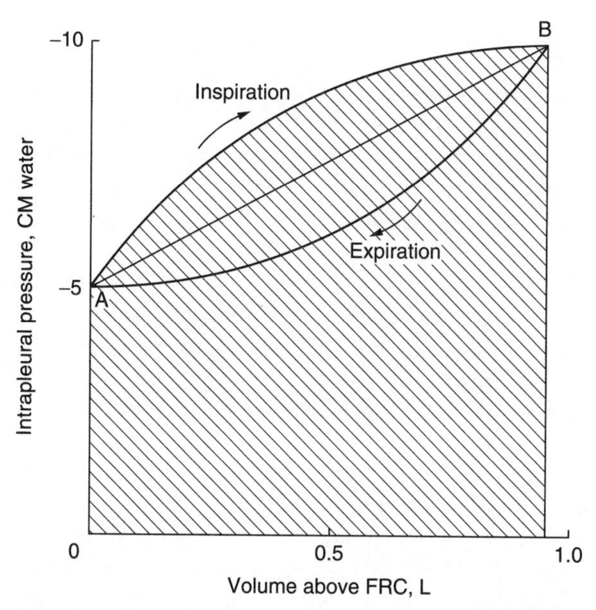

Figure 3.6 Pressure-volume curve showing the total work done by diaphragm due to inspiration.

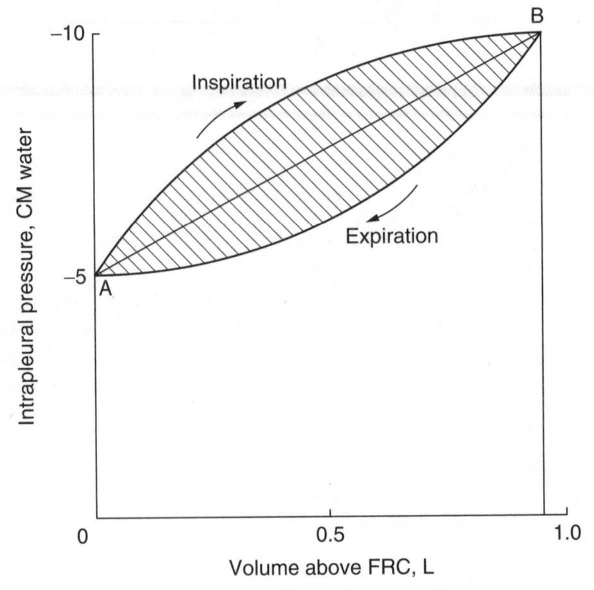

Figure 3.7 Work done by the diaphragm during one breath cycle.

including inspiration and expiration, can be shown as the cross-hatched area in Fig. 3.7.

Clinical features. An important cause of respiratory failure is fatigue of the muscles of respiration. When the diaphragm and respiratory muscles cannot carry out the work of breathing the result is a progressive fall in oxygenation and/or rise in carbon dioxide concentration.

3.7 Airway Resistance

Although the air that flows through your trachea is not very massive or very viscous, there is a noticeable hydraulic resistance to the flow. This resistance results in a pressure drop along the airway. This pressure decreases along the airways, in the direction of flow. This pressure drop is also dependent on the flow rate in the tube, the viscosity of the fluid, and the pattern of flow. There can be no flow along a tube unless there is a pressure difference, or pressure gradient, along the tube.

As explained in Sec. 1.4.1 fluid particles move along streamlines during laminar flow. When air flows at low rates in relatively small diameter tubes, as in the terminal bronchioles, the flow is laminar. Turbulent flow is a random mixing flow. When air flows at higher rates in larger diameter tubes, like the trachea, the flow is often turbulent.

In Chap. 1, we also saw that a dimensionless parameter, termed the Reynolds number, *Re*, could be used to predict whether flow is turbulent or laminar. The number is defined in Eq. (3.7):

$$\mathrm{Re} = \frac{\rho V D}{\mu} \tag{3.7}$$

In Eq. (3.7) ρ is fluid density in kg/m^3, V is fluid velocity in m/s, D is pipe diameter in m, and μ is fluid viscosity in Ns/m^2. Physically the Reynolds number represents the ratio of inertial forces to viscous forces.

As an analogy, imagine the ratio of the momentum of a vehicle (mass × velocity) to the frictional braking force available (force). A person walking on dry pavement has a relatively small mass, relatively low velocity, and relatively low ratio of momentum to stopping force. For comparison, imagine a very large truck moving at high velocity on an icy street. This second combination of truck on ice has a very high ratio of momentum to stopping force and is a highly unstable situation in comparison to the first. Analogously to the person walking on pavement, low mass air flows with low Reynolds numbers are more stable and more likely to be laminar in comparison to denser flows at high velocity with higher Reynolds numbers.

In the lungs, fully developed laminar flow probably occurs only in very small airways with low Reynolds number. Flow in the trachea may be truly turbulent. Much of the flow in intermediate sized airways will be transitional flow in which it is difficult to predict if the flow will be laminar or turbulent.

In Chap. 1 we also developed Poiseuille's law, which describes laminar flow in rigid tubes. Poiseuille's law applies to air flow, just as it does to blood flow, when the flow is laminar. Recall that:

$$Q = \frac{-\pi R^4}{8\mu}\frac{dP}{dx} \tag{3.8}$$

where Q = the flow rate
R = the airway radius
μ = the air viscosity
dP/dx = the pressure gradient along the airway

Resistance to flow can be thought of as the driving pressure divided by the flow. Poiseuille's law can be rearranged to solve for the resistance as shown in Eq. (3.9).

$$\text{Resistance} = \frac{8\mu}{\pi R^4} \tag{3.9}$$

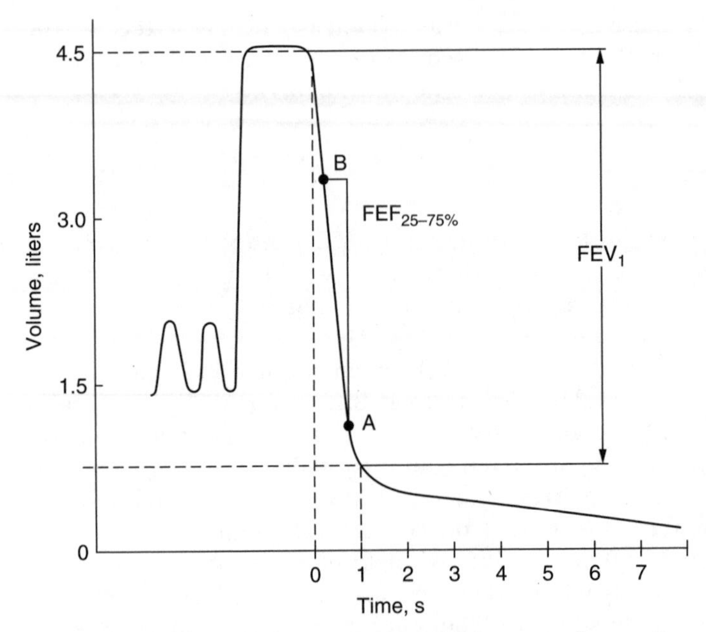

Figure 3.8 Showing a forced expiration versus time curve for a patient with normal airway resistance.

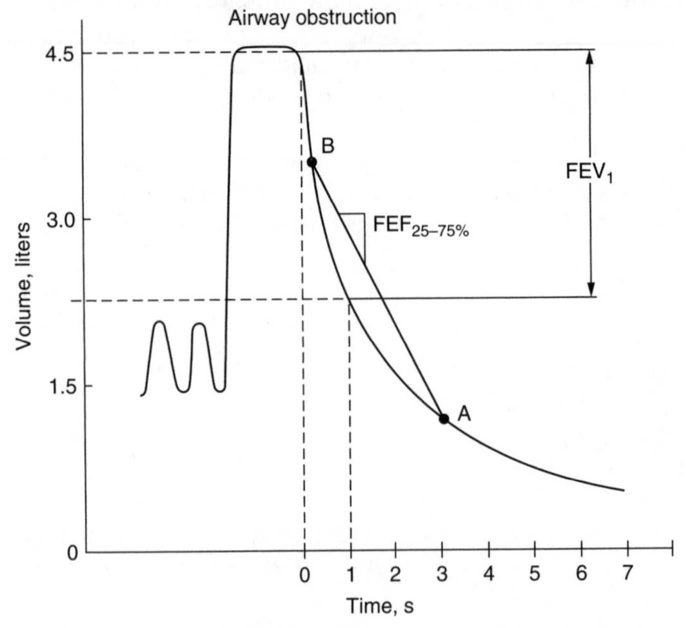

Figure 3.9 The volume versus time curve for forced expiration in a patient with chronic obstructive pulmonary disease shows a great resistance to expiratory flow.

Although the resistance to flow associated with the tube is inversely related to the fourth power of the radius, the resistance contributed by airways is not predominantly in the smallest diameter airways. As airways branch they become narrower, but also more numerous. The major site of airway resistance is the medium–sized bronchi. Small bronchioles contribute relatively little resistance because of their increased numbers.

Most resistance in airways occurs up to the seventh generation of branching vessels. Less than 20 percent of the resistance is attributable to airways that are less than 2 mm in diameter because of the large number of vessels. Twenty-five to 40 percent of total resistance is in the upper airways including the mouth, nose, pharynx, larynx, and trachea.

One way to assess expiratory resistance is to begin by measuring the forced vital capacity (FVC) of the lung, using a spirometer. The patient starts by making a maximal inspiratory effort to total lung capacity. After a short pause, the patient makes a maximal forced expiratory effort.

We may define forced expiratory volume (FEV_1), as the volume of air expired in 1 s during a forced expiratory effort. The ratio of FEV_1 to FVC is a good index of airway resistance. In normal, healthy subjects, FEV_1/FVC is greater than 0.80, or 80 percent.

Forced expiratory flow rate between time t_1 and time t_2 can be defined as the average flow rate between times t_1 and t_2, as measured during a forced expiration. The rate can be represented graphically as the slope of a line drawn between two points on the forced expiration curve.

In Fig. 3.8, point A represents the point on when the lung volume is equal to 25 percent of the vital capacity. A second point, B, is drawn on the curve representing the point when lung volume is equal to 75 percent of vital capacity. The slope of the line AB is the forced expiratory flow rate, $FEF_{25-75\%}$.

Figure 3.8 represents a normal, healthy patient and Fig. 3.9 represents a patient with an airway obstruction. Note the steep slope corresponding to a high forced expiratory flow rate in the healthy patient in Fig. 3.8 and the shallow slope corresponding to a low forced expiratory flow rate in Fig. 3.9.

3.8 Gas Exchange and Transport

In the next section of this book, we will consider how oxygen moves from ambient air into the tissues of the body. Diffusion of a gas occurs when there is a net movement of molecules from an area with a high partial pressure to an area with a lower partial pressure. Only 50 years ago, it was still believed by some scientists that the lung secreted oxygen into the capillaries. That would mean that oxygen would move from the atmosphere to a relatively higher concentration inside the lung by an active process. More accurate measurements have now shown that gas transport across the alveolar wall is a passive process.

3.8.1 Diffusion

Gases like oxygen and carbon dioxide move across the blood-gas barrier of the alveolar wall by diffusion. You might ask, "What parameters affect the rate of transfer?" The rate of gas movement across the alveolar wall is dependent on the diffusion area, the driving pressure, and the wall thickness.

The diffusion area is the surface area of the alveolar wall, or blood-gas barrier. That surface area is proportional to the rate of diffusion. The driving pressure that pushes gasses across the alveolar wall is the partial pressure of the gas in question. The driving force that pushes oxygen into the blood stream is the difference between the partial pressure of oxygen in the alveoli and the partial pressure of oxygen in the blood, DPO_2. The rate of diffusion of oxygen into blood stream is proportional to DPO_2.

The rate of diffusion of oxygen into the blood stream is inversely proportional to the thickness of the alveolar wall. A thicker wall causes the oxygen diffusion to decrease and a thinner wall makes it easier for oxygen to flow into the blood stream.

The blood gas barrier in the lung has a surface area of about 50 to 100 m^2 spread out over 750 million alveoli. This huge diffusion area is available to a relatively small volume of blood. Pulmonary capillary blood volume is only about 60 mL during resting and about 95 mL during exercise.

Carbon dioxide diffuses about 20 times more rapidly than oxygen through the alveolar wall. The much higher solubility of carbon dioxide is responsible for this increased diffusion rate.

3.8.2 Diffusing capacity

One might wonder if oxygen transfer into the blood is limited by how fast the blood can flow through the lungs, or by how fast oxygen can diffuse through the blood-gas barrier. It turns out that under normal physiological circumstances, oxygen diffusion through the alveolar wall is sufficient and the limiting factor for oxygen uptake is perfusion, or the rate of blood flow through the capillaries. An erythrocyte spends an average of about 0.75 to 1.2 s passing through a pulmonary capillary under normal resting conditions.

Under some circumstances oxygen diffusion rates through the wall may also limit the oxygen transfer rate. During extreme exercise, an erythrocyte may stay in the capillary as little as 0.25 s. Even this very short time would be enough time for oxygen to diffuse into the capillaries at normal atmospheric oxygen partial pressures. However, during extreme exercise at high altitude, or in a patient with thickening of the alveolar wall due to pulmonary fibrosis, oxygen flow rate may switch from a perfusion limited to a diffusion limited process.

3.8.3 Resistance to diffusion

So far, we have considered the blood gas barrier as the only source of resistance to diffusion of gases from the alveoli into the blood stream. In fact, there is a second important component. The uptake of oxygen occurs in two stages. The first is diffusion through the alveolar wall, and the second is the reaction of oxygen with hemoglobin.

The diffusing capacity through the alveolar wall is defined by the flow rate of the gas divided by the partial pressure difference that is driving the flow. Diffusing capacity of a gas can be written as D_a in Eq. (3.10).

$$D_a = \frac{\dot{V}_{\text{gas}}}{(P_a - P_c)} \tag{3.10}$$

In Eq. (3.10), $\dot{V}_{\text{gas}}$ represents the flow rate of some gas from the alveoli into the capillary. P_a represents the partial pressure of that gas in the alveoli and P_c represents the partial pressure of the same gas in the capillary.

If we think of the electrical analogy in which resistance is equal to voltage divided by current, we can see that it is possible to think of $1/D_a$ as a resistance to diffusion.

$$\frac{1}{D_a} = \frac{P_a - P_c}{\dot{V}_{\text{gas}}} \tag{3.11}$$

The rate of reaction of oxygen with hemoglobin can be represented by θ. The units on θ are ml O_2/min per mL blood per mmHg partial pressure of O_2. If Vc represents the volume of pulmonary capillary blood, then $\theta \times V_c$ will have the units of mL O_2/min/mmHg, or flow divided by pressure. Once again, the inverse, pressure/flow is analogous to resistance.

$$\text{Resistance}_{\text{O}_2\text{-Hgb}} = \frac{1}{\theta V_c} \tag{3.12}$$

The total resistance to the flow of oxygen into the blood stream is the combination of the resistance to diffusion caused by the blood gas barrier plus the resistance to flow due to the oxygen-hemoglobin reaction.

The resistances can be added together as shown in Eq. (3.13). $1/D_L$ is the total resistance to the flow of oxygen due to the lung.

$$\frac{1}{D_L} = \frac{1}{D_a} + \frac{1}{\theta V_c} \tag{3.13}$$

For oxygen flow, the resistance to diffusion offered by the membrane is approximately equal to the resistance associated with the oxygen-hemoglobin reaction. It is also interesting to note that carbon dioxide diffusion is approximately twenty times faster than oxygen diffusion. Therefore it seems unlikely that CO_2 elimination will be slowed by an increased resistance to diffusion.

3.8.4 Oxygen dissociation curve

One gram of pure hemoglobin can combine with 1.39 mL of oxygen and normal human blood has approximately 15 g of hemoglobin in each 100 mL of whole blood. This means that the oxygen carrying capacity of blood is about 20.8 mL oxygen/100 mL blood.

Oxygen saturation is reported as a percent and is equal to the amount of oxygen combined with hemoglobin at a given partial pressure divided by the maximum oxygen carrying capacity. For example, oxygen saturation of blood is about 75 percent when the blood is exposed to air with a partial pressure for oxygen of 40 mmHg and 97.5 percent saturation at a partial pressure of oxygen of 100 mmHg. Figure 3.10 shows a normal oxygen dissociation curve from a human.

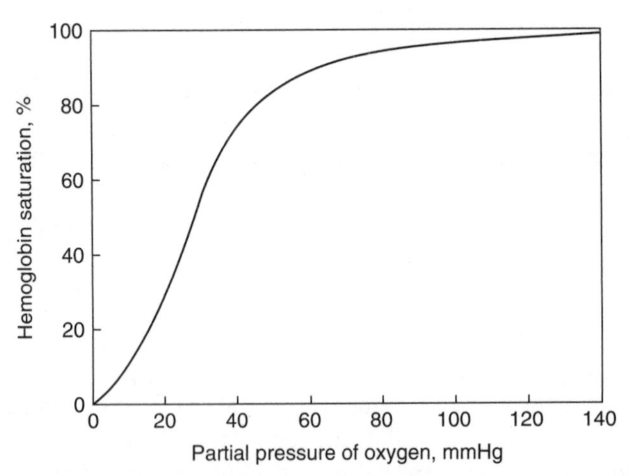

Figure 3.10 Oxyhemoglobin dissociation curve.

Oxygen dissolved in blood, rather than carried in the hemoglobin, amounts to 0.003 mL O_2 per 100 mL blood per mmHg partial pressure. Therefore, the amount of dissolved oxygen in blood with a driving pressure of oxygen equal to 100 mmHg is only 0.3 mL O_2 per 100 mL blood, compared to 20.3 mL O_2 carried in the hemoglobin.

3.9 Pulmonary Pathophysiology

3.9.1 Bronchitis

Bronchitis is an inflammation of the airways resulting in excessive mucus production in the bronchial tree. Bronchitis occurs when the inner walls of the bronchi become inflamed. It often follows a cold or other respiratory infection and happens in virtually all people, just as the common cold. When the bronchitis does not go away quickly but persists, then it is termed chronic bronchitis.

3.9.2 Emphysema

Emphysema is a chronic disease, in which air spaces beyond bronchioles are increased. See Fig. 3.11. The stiffness of the alveoli is decreased (static compliance is increased) and airways collapse more easily. Because of the decreased stiffness of the lung, exhalation requires active work and the work of breathing is significantly increased. The surface area of the alveoli become smaller, and the air sacs become less elastic. As carbon dioxide accumulates in the lungs, there becomes less and less room available for oxygen to be inhaled, thereby decreasing the partial pressure of oxygen in the lung.

Emphysema is most often caused by cigarette smoking although some genetic diseases can cause similar damage to the alveoli. Once this damage has occurred, it is not reversible.

3.9.3 Asthma

Asthma is a chronic disease that currently affects 5 million children in the United States. In asthma, the airways become over reactive with increased mucus production, swelling, and muscle contraction. Because of the decreased size of the bronchi and bronchioles, flow of air is restricted and both inspiration and expiration become more difficult.

3.9.4 Pulmonary fibrosis

Pulmonary fibrosis currently affects 5 million people worldwide and 200,000 in the United States. Pulmonary fibrosis is caused by a thickening or scarring of pulmonary membrane. The result is that the alveoli are

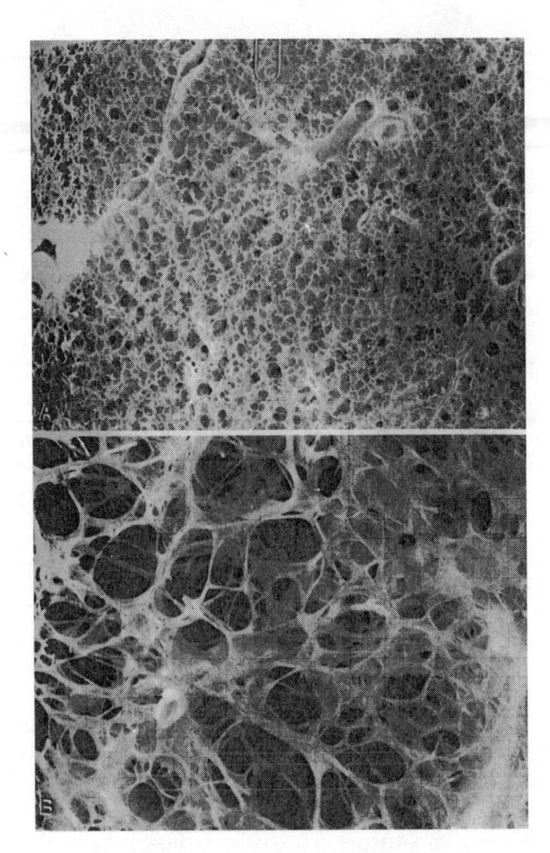

Figure 3.11 shows a normal lung and an emphysematous lung. (a) Normal lung. (b) Emphysematous lung. (Reprinted with permission from West J B. *Pulmonary Pathophysiology*. Williams & Wilkins; 4th edition,1992.)

gradually replaced by fibrotic tissue becoming thicker, with a decreased compliance (increased stiffness) and a decrease in diffusing capacity. Symptoms of pulmonary fibrosis include a shortness of breath, chronic dry, hacking cough, fatigue and weakness, chest discomfort, loss of appetite, and rapid weight loss. Traditionally, it was thought that pulmonary fibrosis might be an autoimmune disorder or the result of a viral infection. There is growing evidence that there is a genetic link to pulmonary fibrosis.

3.9.5 Chronic obstructive pulmonary disease

Chronic obstructive pulmonary disease (COPD) is a slowly progressive disease of the lung and airways. COPD can include asthma, chronic bronchitis, chronic emphysema, or some combination of these conditions. The disease is characterized by a gradual loss of lung function. The most significant risk factor for COPD is cigarette smoking. Other documented causes of COPD include occupational dusts and chemicals. Genetic factors can also play a significant role in some forms of this disease.

3.9.6 Heart disease

Heart disease should be mentioned in any discussion of pulmonary pathologies. While cardiac disease is not strictly speaking, a pulmonary pathology, some forms of cardiac disease can certainly lead to respiratory pathologies. For example, a stenotic regurgitant mitral valve can cause back pressure in pulmonary capillary leading to fluid in the lungs.

3.9.7 Comparison of pulmonary pathologies

Figure 3.12 shows the volume versus time curve for a normal lung compared to that for a patient with fibrosis, asthma, and emphysema. In the patients with fibrosis, asthma, and emphysema note the shallow slope of the curve during forced inspiration and expiration. In other words, the change in volume over time dV/dt is much smaller in patients with lung disease compared to the dV/dt in the normal lung. The air flow rate is much smaller in all three cases.

3.10 Respiration in Extreme Environments

Consider how you might feel if you drive your automobile to the top of Pike's Peak (14,109 ft above sea level) or if you ride a cable car to the top of the Zugspitze, the highest point in Germany (9718 ft above sea level). If you have had the opportunity to visit either of these locations, you probably experienced the shortness of breath associated with breathing in low oxygen environments. Perhaps you even developed a headache after a short period. How you felt was dependent on how long it took to achieve the altitude, how long you remained, how well hydrated you may have been at the time, and a number of other potential factors.

3.10.1 Barometric pressure

Just as with normal respiration, at high altitude the driving force that helps to push oxygen into your blood is the partial pressure of oxygen. This partial pressure depends on both the barometric pressure and the relative percentage of air that consists of oxygen. Barometric pressure depends on the altitude above the earth's surface and varies approximately exponentially as shown in Fig. 3.13.

The equation for barometric pressure as a function of altitude depends on the density of the air at varying altitudes and therefore on air temperature. The equation for the standard atmosphere between sea level and 11 km above the earth's surface can be given by:

$$P_{\text{atm}} = 760(1 - 0.0022558z)^{5.2559}$$

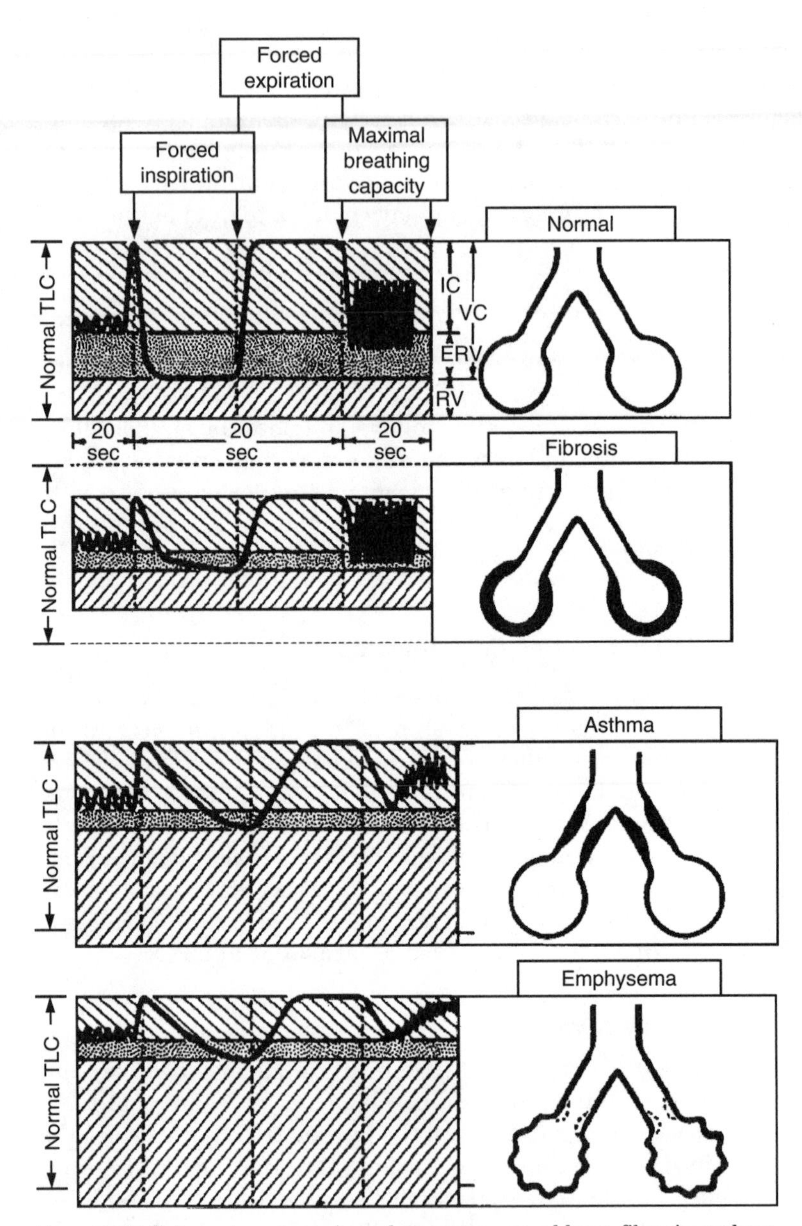

Figure 3.12 Spirometer comparisons between a normal lung, fibrosis, asthma, and emphysema. (Reprinted with permission of Lippincott, Williams, and Wilkins from Selkurt, *Basic Physiology for the Health Sciences*, Little, Brown and Company, Boston, second edition, 1982.)

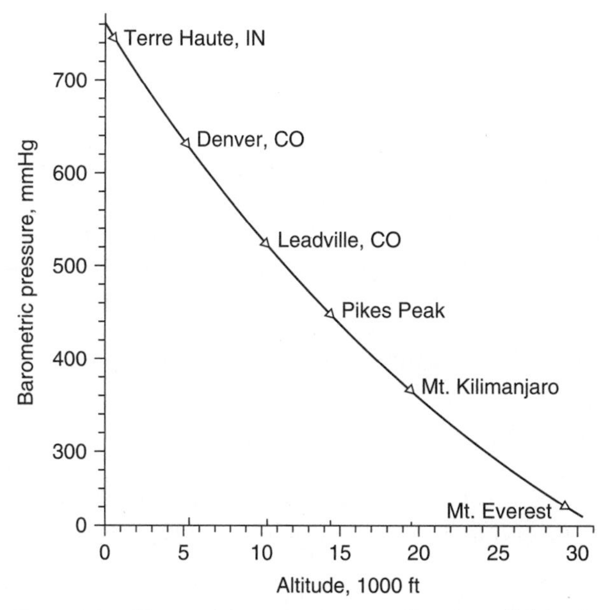

Figure 3.13 Barometric pressures as a function of altitude above sea level.

where P_{atm} is the barometric pressure in millimeters of mercury, and z is the altitude above mean sea level in kilometers.

From the figure it is possible to see that on a standard day, the barometric pressure would decrease from 747 mmHg in Terre Haute, Indiana, to 625 mmHg in Denver, Colorado, to 360 mmHg at the summit of Mt. Kilimanjaro and 235 mmHg at the summit of Mt. Everest. With the decrease in altitude and corresponding decrease in barometric pressure comes a decrease in the partial pressure of oxygen.

3.10.2 Partial pressure of oxygen

The partial pressure of oxygen, PO_2, is the driving pressure for getting oxygen into the blood. Air is 21 percent oxygen, so the partial pressure of oxygen in standard air is 0.21 times the barometric pressure. The PO_2 in air on a standard day in Terre Haute, Indiana is $0.21 \times 747 = 157$ mmHg. If PO_2 is low, the driving pressure to push oxygen into the blood stream will be low making it more difficult to breath.

Further, the air inside your lungs is not dry air. Water vapor also displaces oxygen. In fact, the air in your lungs is saturated with water, and the vapor pressure of water at 37 °C, the temperature inside your lungs, is 47 mmHg. Now the PO_2 of sea level, dry air is

$$760\ (0.2093) = 159\ \text{mmHg},$$

but the PO_2 of saturated, inspired air is

$$(760 - 47)(0.2093) = 149 \text{ mmHg}.$$

If you climb even higher, the driving force becomes lower but the vapor pressure of water does not change since the air in your lungs is always saturated (100 percent relative humidity).

Let's climb to 14,000 ft above sea level (ASL). Now the PO_2 of inspired air at 14,000 ft ASL is $(425 - 47)(0.2093) = 79$ mmHg or about 1/2 of the PO_2 for inspired air at sea level.

Now if we continue to climb to 18,000 ft above sea level the PO_2 in dry air is half of that of sea level dry air and the inspired partial pressure of oxygen is only 70 mmHg.

$$(380 - 47)(0.2093) = 70 \text{ mmHg}$$

Finally, at the top of Mt. Everest the PO_2 is one-third of sea level dry air and the PO_2 of inspired air at 29,000 ft above sea level is only 43 mmHg.

$$(250 - 47)(0.2093) = 43 \text{ mmHg}$$

What would happen if a high-performance jet lost cabin pressure at an altitude of 63,000 ft above sea level? The barometric pressure at this extreme altitude is less than the vapor pressure of water at 37 °C. In this case the partial pressure of oxygen would be zero and in fact, your blood (and all other water in your tissues) would boil.

3.10.3 Hyperventilation

Inspired oxygen is not the complete story. Consider the CO2 in your lungs. Doesn't it also displace air, making the PO_2 even lower? The partial pressure of carbon dioxide in your lungs can approach 40 mmHg. Since the partial pressure of inspired oxygen at an altitude of 18,000 ft above sea level was calculated as 70 mmHg without considering CO_2, the PO_2 of air in the alveoli could be as low as $70 - 40 = 25$ mmHg. At the top of Mt. Everest it could be as low at $43 - 40 = 3$ mmHg. If that is true, how can so many people climb above 14,000 ft so easily and a few people even reach the summit of Everest without oxygen? The short answer is hyperventilation. By breathing faster, climbers are able to lower the partial pressure of carbon dioxide in their alveoli. If you increase ventilation rate by 4, you can lower the PCO_2 to about 10 mmHg.

With hyperventilation, the PO_2 of alveolar oxygen at the top of Mt. Everest can be calculated as follows.

The partial pressure of oxygen contained in the inspired air that one would breathe at an altitude of 29,000 feet is $(250 - 47)(0.2093) = 43$ mmHg. The PCO_2 of the air in the alveoli is 10 mmHg. The partial pressure of the oxygen in the alveoli is $43 - 10 = 33$ mmHg.

3.10.4 Alkalosis

One might wonder whether there are any difficulties or side effects resulting from hyperventilation and the associated drop in the partial pressure of carbon dioxide. In fact, the result is respiratory alkalosis. The pH of the blood increases above normal. You feel bad and you cannot sleep well. One result could be acute mountain sickness.

Your body's solution to respiratory alkalosis is that your kidneys will excrete bicarbonate over the next few days. At the same time, the increase in pH puts a kind of "brake" on ventilation and causes your breathing to slow down. The excretion of bicarbonate causes the pH in the blood to decrease and the "brakes" on ventilation are reduced. After two or three days at altitude, blood pH returns to normal.

3.10.5 Acute mountain sickness (AMS)

The feeling of nausea, and headaches resulting from low oxygen environments, hyperventilation, and the associated respiratory alkalosis, is known as acute mountain sickness. Sleeping at altitude is also troublesome. When your breathing slows, oxygen saturation in your blood drops. This drop in PO_2 causes you to wake up with a feeling of breathlessness. Diamox is a trademark name for acetazolamide, a well-known carbonic anhydrase inhibitor, which is used in the treatment of seizures and glaucoma and is also a diuretic. Diamox does not prevent mountain sickness, but it speeds up acclimatization by increasing urinary excretion of bicarbonate. People taking Diamox have been shown to have a more consistent level of blood oxygen saturation, enabling more restful sleep. Acute mountain sickness can also result in nocturnal periodic breathing, weird dreams, and frequent awakening at night.

AMS has been recognized for many centuries. Great Headache Pass and Little Headache Pass are the names of two Himalayan passes between China and Afghanistan that were named in ancient times. Also, Father Joseph Acosta, a Jesuit Priest who lived in Peru in the sixteenth century, described AMS and deaths, which occurred high in the Andes.

3.10.6 High-altitude pulmonary edema

High-altitude pulmonary edema (HAPE) is a life-threatening noncardiogenic (not caused by heart disease) lung edema. The mechanism that causes HAPE is not completely known, but the disease is thought to be caused by patchy, low-oxygen, pulmonary vasoconstriction (constriction of blood vessels). That constriction results in localized over perfusion and increased permeability of pulmonary capillary walls. These changes result in high pulmonary artery pressure, high permeability, and fluid leakage into the alveoli. The result of high-altitude pulmonary edema is that the lungs fill with fluid, decreasing ventilation. Symptoms of HAPE can include, cough, shortness of breath on exercise, progressive shortness of breath, and eventually suffocation if left untreated. The condition is unstable and the only effective treatment for HAPE is descent to a lower altitude and respiration in a more oxygen rich environment.

3.10.7 High-altitude cerebral edema

High-altitude cerebral edema (HACE) is a less common, but equally life threatening condition in which cerebral edema is the result of breathing in a depleted oxygen environment. HACE is theoretically linked to brain swelling. Symptoms of HACE can include severe throbbing headache, confusion, difficulty walking, difficulty speaking, drowsiness, nausea, vomiting, seizures, hallucinations, and coma. A person suffering from HACE is often gray or pale in appearance. A victim can suffer from HAPE and HACE simultaneously. Both HAPE and HACE normally occur at altitudes above 15,000 ft above sea level but can occur at high ski area above 8000 ft above sea level.

3.10.8 Acclimatization

If you ascend to a high altitude and remain for a long period of time, your body can begin to acclimatize. By increasing the number of red blood cells in your blood, you are able to increase the amount of hemoglobin and hence the oxygen carrying capacity of your blood. There are people who live permanently above 16,000 ft in the Peruvian Andes. Their inspired PO_2 is only about 45 mmHg, but those people have more oxygen in their blood than do normal sea level residents!

The volume percent of red blood cells in blood is known as hematocrit. Hematocrit in normal healthy males is about 42 to 45 percent. If a person's hematocrit falls below 25 percent, then he or she is considered anemic. Hematocrit values can also be greater than normal and a hematocrit above 70 percent is known as polycythemia.

During acclimatization to low oxygen environments, it is possible to replace approximately 1 percent of your erythrocytes per day. At the same time, erythrocytes continue to die after their approximately 125-day life span. Therefore, it takes two or three weeks to increase your hematocrit significantly.

In a study done in Potosi, Bolivia, which is located in southern Bolivia on a high plane at approximately 13,000 ft above sea level, the average hematocrit in males was found to be 52.7 percent. Early observations of human adaptation to high altitude were made in Potosi by de la Calancha in the late 1600s. The Spanish historian writes that "more than one generation was required" for a Spanish child to thrive in Potosi. It is not clear whether being born and raised at high altitude or whether interbreeding with the Andean population was required.

3.10.9 Drugs stimulating red blood cell production

Erythropoietin (EPO) is a naturally occurring hormone which stimulates red blood cell production. Low oxygen stimulates EPO production in humans, which in turn stimulates a higher production of red blood cells. Recombinant EPO (rEPO) is a synthetic version of this hormone.

The difficulty with taking rEPO to increase hematocrit is that an increase in hematocrit also increases blood viscosity, and therefore increases the work of the heart required to pump blood. At the 1984 Los Angeles Olympics before rEPO was available; some American bicycle racers received blood transfusions to raise their red blood cell counts (they also received hepatitis). After rEPO became readily available athletes began taking injections of rEPO. A number of Belgian and Dutch professional cyclists died of strokes in 1987–88, presumably from erythropoietin-induced clots in arteries.

Recombinant human erythropoietin (rEPO) has become the standard of care for renal anemia. EPO has a relatively short half-life and is generally administered two or three times a week.

Darbepoetin is a synthetic hormone that increases red blood cell production, and is used to treat anemia and related conditions. It is in the same class of drugs as recombinant erythropoietin (rEPO) and competes for the same market. Its brand name is Aranesp, and it is marketed by Amgen. It was approved in September 2001 by the U.S. Food and Drug Administration for treatment of patients with chronic renal failure by intravenous or subcutaneous injection. Darbepoetin alpha is a longer lasting agent with a half-life approximately three times that of rEPO.

Like rEPO, Darbepoetin's use increases the risk of cardiovascular problems, including cardiac arrest, seizures, arrhythmia or strokes, hypertension, congestive heart failure, vascular thrombosis, myocardial infarction, and edema. Also like rEPO, Darbepoetin has the potential to be abused by athletes seeking an advantage. Its use during the 2002 Winter Olympic Games to improve performance led to the disqualification of several cross-country skiers from their final races.

Bibliography

Selkurt E. *Basic Physiology for the Health Sciences*, second edition. Boston, MA: Little, Brown & Co; 1982.

Wark K Jr., Richards DE. *Thermodynamics, sixth edition*. New York: McGraw-Hill; 1999.

West J B. *Respiratory Physiology*. Baltimore,Williams & Wilkins; 1990.

West J B. *Pulmonary Pathophysiology*. Baltimore, Williams & Wilkins; 1992.

Chapter

4

Hematology and Blood Rheology

4.1 Introduction

Rheology is the study of the deformation and flow of matter. Chapter 4 presents the study of blood and especially the properties that are associated with the deformation and flow of blood.

4.2 Elements of Blood

Blood consists of 40 to 45 percent formed elements. Those formed elements include red blood cells or erythrocytes, white blood cells or leukocytes, and platelets or thrombocytes. Erythrocytes are those cells that are involved primarily in the transport of oxygen and carbon dioxide. Leukocytes are cells that are involved primarily in phagocytosis and immune responses, while thrombocytes are involved in blood clotting.

In addition to the formed elements in blood 45 to 60 percent of blood by volume consists of plasma. Plasma is the transparent, amber colored liquid in which the cellular components of blood are suspended. It also contains things like proteins, electrolytes, hormones, and nutrients. Serum is blood plasma from which clotting factors have been removed.

4.3 Blood Characteristics

Blood accounts for 6 to 8 percent of body weight in normal, healthy humans. For example, a male with a body weight of 165 lb can expect to have $165 \times (0.07) = 12$ lb of blood. The density of blood is slightly larger than the density of water at approximately 1060 kg/m^3. The increased density comes from the increased density of a red blood cell compared to the density of water or plasma. The density of water is 1000 kg/m^3.

Therefore, those 12 lb of blood have a volume of about 12 pt. Most people have between 4.5 and 6 L of blood.

In Sec. 1.2.2, we learned about a fluid property known as viscosity. Viscosity is defined by the slope of the curve on a shear stress versus shearing rate diagram. Viscosity of the blood is one of the characteristics of blood that affects the work required to cause the blood to flow through the arteries. The viscosity of blood is in the range of 3 to 4 cP, or 0.003 to 0.004 Ns/m^2. For comparison, the viscosity of water at room temperature is approximately 0.7 cP. Blood is a non-newtonian fluid, which we learned, means that the viscosity of blood is not a constant with respect to the rate of shearing strain.

In addition to the rate of shearing strain, the viscosity of blood is also dependent on temperature and the volume percentage of blood that consists of red blood cells.

The term hematocrit is defined as the volume percent of blood that is occupied by erythrocytes, or red blood cells. Since erythrocytes are the cells that make up the oxygen and carbon dioxide carrying capacity of blood, the hematocrit is an important parameter affecting the bloods ability to transport these gases. A normal hematocrit in human males is 42 to 45 percent. Hematocrits under 40 percent are associated with anemia. Hematocrits above 50 percent are associated with a condition called polycythemia in which the number of red blood cells in an individual is increased above normal.

4.4 Erythrocytes

The name erythrocyte comes from the Greek *erythros* for red and *kytos* for hollow, which is commonly translated as cell. Erythrocytes are biconcave discs with a diameter of approximate 8 μm. Figure 4.1 shows a drawing of a typical erythrocyte showing the basic geometry. Figure 4.2 shows a photomicrograph of an erythrocyte. The volume of the typical erythrocyte is approximately 85 to 90 mm^3. The shape of an erythrocyte gives a very large ratio of surface area to volume for the cell. In fact, a sphere with the same volume of a typical red blood cell would have only 60 percent as much surface area.

The life span of a red blood cell is approximately 125 days. This means that about 0.8 percent of all of your red blood cells are destroyed each day, while the same amount of red blood cells are also produced in the bone marrow.

Erythrocyte formation is known as erythropoiesis and occurs in the red marrow of many bones including the long bones like the humerus, and femur as well as other types of bones like the skull and ribs. Iron is required to produce the hemoglobin necessary for red blood cell production and one to four milligrams of iron per day is the minimum

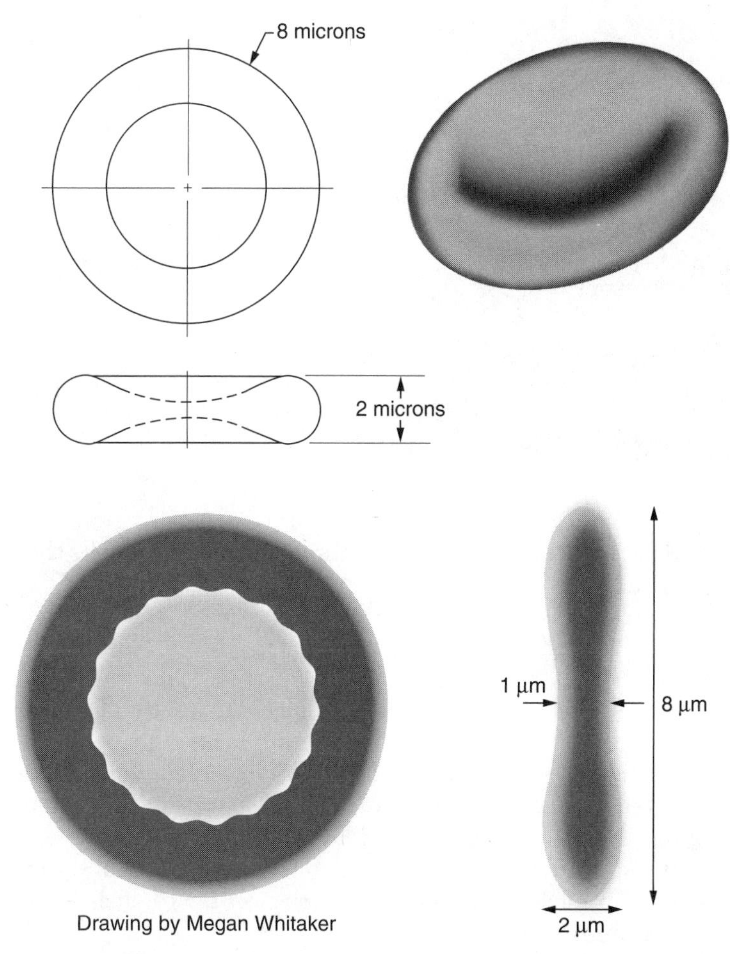

Figure 4.1 Dimensions of an erythrocyte.

required. Erythropoietin stimulates erythrocyte production in response to hypoxia.

Erythrocyte destruction—erythrocytes live for approximately 125 days. At the end of a red blood cell's life span, the erythrocyte becomes fragile and disintegrates. Erythrocytes are destroyed by the macrophages of the mononuclear phagocytic system. The iron is taken to the bone marrow where it is recycled into new hemoglobin.

Normal, healthy human blood typically contains approximate 5 million red blood cells in each cubic millimeter of whole blood. This number depends, of course, on blood hematocrit as well as the size of the red blood cells. Mature mammalian erythrocytes do not have a nucleus and contain no RNA, no Golgi apparatus, and no mitochondria. Despite the

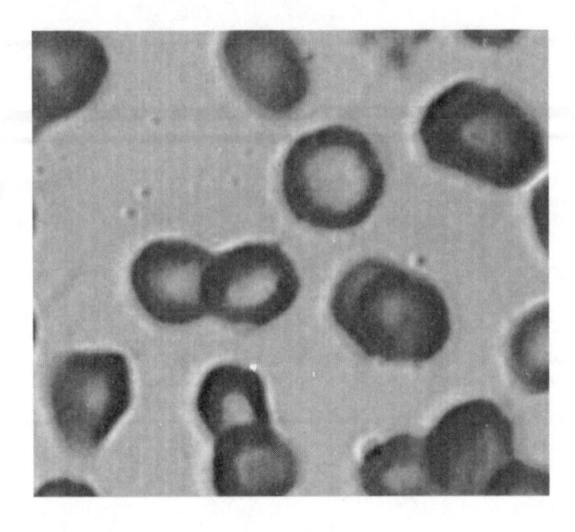

Figure 4.2 Photograph of erythrocytes.

fact that they do not have mitochondria, erythrocytes are living cells that are metabolically active.

4.4.1 Hemoglobin

Hemoglobin (Hgb) is a large molecule with a molecular weight of 67,000 daltons (Da). A dalton is equivalent to one atomic mass unit, which is equal to one-twelfth of the mass of a carbon 12 molecule. Hemoglobin is also a conjugated protein, which means that it is a protein with an attached nonprotein portion. The hemoglobin molecule is composed of four polypeptide subunits (the protein or globin components) and four heme groups (one on each globin chain). At the center of each heme group is an iron atom (Fe^{2+}).

The sigmoid shape of the oxyhemoglobin dissociation curve is a consequence of the four subunits of hemoglobin cooperating in the binding of oxygen. At low oxygen pressures, there is a low probability that one or more than one of the four subunits will have an oxygen molecule bound to it. As the pressure increases and oxygen concentration increases, there is an increasing probability that at least one subunit has an oxygen bound. Binding of oxygen to one of the subunits increases the probability that the other subunits will be able to bind an oxygen molecule. So, as oxygen pressure increases even further, the probability that more and more of the remaining binding sites will have an oxygen molecule bound to them rapidly increases.

Hemoglobin readily associates and disassociates with/from oxygen and carbon dioxide. Hemoglobin that is associated with oxygen is known as oxyhemoglobin and is bright red in color. Reduced hemoglobin that has been disassociated from oxygen has a bluish purple color. The heme submolecule of the hemoglobin is the portion of Hgb that associates with oxygen. Each hemoglobin molecule can transport four O_2 molecules.

The globin submolecule is the protein portion of hemoglobin that associates with carbon dioxide. Each globin sub-molecule can transport one molecule of CO_2. Normal adult hemoglobin includes three types. These three types of hemoglobin include type A hemoglobin, type A_2 hemoglobin, and type F hemoglobin. Type A makes up 95 to 98 percent of hemoglobin in adults. Type A_2 makes up 2 to 3 percent and type F, fetal hemoglobin makes up a total of 2 to 3 percent. Fetal hemoglobin is the primary hemoglobin produced by the fetus during gestation and it falls to a much lower level after birth.

S type hemoglobin is associated with sickle cell anemia. Hemoglobin S forms crystals when exposed to low oxygen tension. Sickle cells are fragile and hemolyze easily. When the erythrocytes sickle, become fragile, and hemolyze they could block the microcirculation, causing infarcts of various organs. Figure 4.3 shows sickle cell erythrocytes. Persons with S type hemoglobin are also resistant to malaria which helps this type of hemoglobin to persist in the populations where malaria in endemic.

4.4.2 Clinical Features

Sickle cell anemia turns up as severe anemia caused by hemolysis (breaking up of red blood cells). This anemia is punctuated by crises. The symptoms of the anemia are mild in comparison to the severity of the anemia because S type hemoglobin gives up oxygen relatively easily compared to hemoglobin A. Some patients have an almost normal life, with few crises. Other patients develop severe crises and may die in early childhood. The sickle cell crises are sometime very painful. The crises can be precipitated by such factors as infection, acidosis, dehydration, or deoxygenation (associated with altitude for example). The most serious

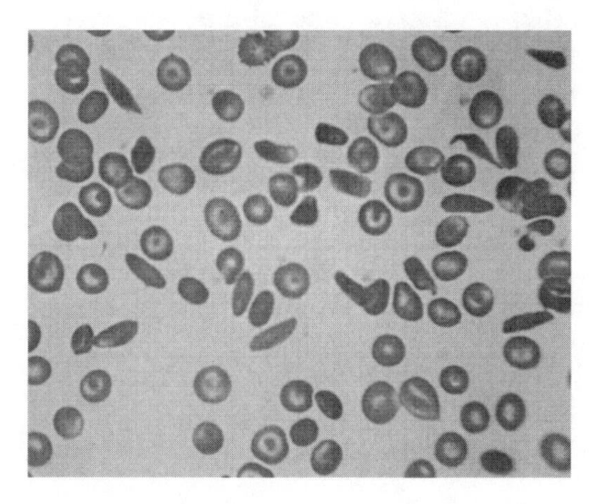

Figure 4.3 Sickle cell erythrocytes.

vascular occlusive crisis occurs in the brain. Seven percent of all sickle cell anemia patients experience a stroke.

4.4.3 Erythrocyte indices

There can be a number of reasons to cause a person to have a lower than normal quantity of hemoglobin. It could be that the person has a normal number of erythrocytes but lower than normal hemoglobin in each erythrocyte. It could be that the person has a normal amount of hemoglobin in each erythrocyte, but a very low hematocrit. These parameters are quantified and those values are known as erythrocyte indices.

One index measures the average volume of an erythrocyte and it is known as mean corpuscular volume (MCV). The mean corpuscular volume can be calculated from the hematocrit of whole blood and the concentration of red blood cells per unit volume as in Eq. (4.1).

$$\mathrm{MCV} = \frac{\mathrm{HCT}}{\dfrac{\text{No. of RBC}}{\text{volume}}} \tag{4.1}$$

Example If a patient has a hematocrit of 45 percent red blood cells and if that same person has 5 million red blood cells in each cubic millimeter of blood. Estimate the MCV.

$$\mathrm{MCV} = \frac{0.45}{5 \times 10^6 \dfrac{\mathrm{RBC}}{\mathrm{mm}^3} \times \dfrac{1000^3\,\mathrm{mm}^3}{\mathrm{m}^3}} = 90 \times 10^{-18}\,\mathrm{m}^3 = 90\,\mu\mathrm{m}^3$$

$1\ micron = 10^{-6}$ meters

A second index measures the amount of hemoglobin in a red blood cell. That index is known as the mean corpuscular hemoglobin index or MCH. MCH may be calculated as shown in Eq. (4.2).

$$\mathrm{MCH} = \frac{\dfrac{\mathrm{Hgb}}{\text{volume}}}{\dfrac{\text{No. of RBC}}{\text{volume}}} \tag{4.2}$$

Example If a patient has a hemoglobin concentration of 15 g per 100 mL of whole blood and also has five million red blood cells in each mm^3 of blood, then the MCH can be calculated in the following manner:

$$\mathrm{MCH} = \frac{\dfrac{15\,\mathrm{g}}{100\,\mathrm{ml}}}{5 \times 10^6\,\mathrm{RBC/mm}^3} = \frac{\dfrac{15\,\mathrm{g}}{100\,\mathrm{ml}} \times \dfrac{1\,\mathrm{ml}}{\mathrm{cm}^3} \times \dfrac{\mathrm{cm}^3}{10^3\,\mathrm{mm}^3}}{\dfrac{5 \times 10^6\,\mathrm{RBC}}{\mathrm{mm}^3}} = 30 \times 10^{-12}\,\frac{\mathrm{gm}}{\mathrm{RBC}}$$

A third erythrocyte index is mean corpuscular hemoglobin concentration or MCHC. This index is a measure of the concentration of hemoglobin in whole blood (mass/volume). The MCHC may be calculated as shown below in Eq. (4.3).

$$\mathrm{MCHC} = \frac{\frac{\mathrm{Hgb}}{\mathrm{volume}}}{\mathrm{Hematocrit}} \tag{4.3}$$

Example For a patient with a hemoglobin concentration of 15 g per 100 mL of whole blood and a hematocrit of 0.45 the MCHC can be calculated as shown below.

$$\mathrm{MCHC} = \frac{\frac{15\ \mathrm{gm}}{100\ \mathrm{ml}}}{0.45} = \frac{0.33\ \mathrm{gm\ of\ hemoglobin}}{100\ \mathrm{ml\ blood}}$$

4.4.4 Abnormalities of the blood

Anemia is defined as a reduction below normal in the oxygen carrying capacity of blood. This reduction in oxygen carrying capacity could be a result of a decrease in the number of red blood cells in the blood, or it could be a result in a decrease in hemoglobin in each red blood cell, or it could be caused by both.

A macrocytic hyperchromic anemia is an anemia in which a greatly reduced number of red blood cells are too large (macrocytic), have too much hemoglobin and are therefore too red (hyperchromic). These large, red erythrocytes lack erythrocyte maturation factor. This lack can sometime be related to a vitamin B12 deficiency.

Polycythemia is defined as an increase in the number of erythrocytes per cubic millimeter of whole blood above 6 million. Relative polycythemia can be a result of increased concentration of red blood cells due to decreased blood volume, for example as a result of dehydration. Polycythemia vera, however, results from hyperactivity of the bone marrow. It is sometimes difficult to determine the cause of polycythemia vera, but it may be symptomatic of tumors in the bone marrow, kidney, or brain.

4.5 Leukocytes

Leukocytes are also known as white blood cells. Leukocytes can be broadly defined into two groups, arranged by function—phagocytes and immunocytes. They could also be arranged into two groups by appearance—granulocytes and agranulocytes. Healthy whole blood normally contains approximately 4000 to 11,000 leukocytes in each cubic millimeter. If you compare that number to 5 million erythrocytes per cubic millimeter, then one would expect to see around 500 erythrocytes for every leukocyte.

TABLE 4.1 Various Types of Leukocytes Grouped in Order of Their Relative Numbers

Name	Count per mm^3	Size, μ
Neutrophil	2500–7500	10–15
Lymphocyte	1000–3000	10–20
Monocyte	200–800	20–25
Eosinophil	40–400	10–15
Basophil	10–100	10–12
Total leukocytes	4000–11000	

Leukocytes are translucent. If we look at an unstained blood smear under the microscope normally, we will not see any leukocytes. If we searched very carefully and diligently, perhaps we could find what appears as a white blood cell ghost. Leukocytes each contain a nucleus and other organelles and are easy to find after staining.

The two groups of leukocytes based on function are the phagocytes, which spend their time eating foreign bodies, and immunocytes, which are involved in the body's immune response. Phagocytes include neutrophils, eosinophils, monocytes, and basophils. The cells known as immunocytes are lymphocytes (see Table 4.1).

If we divide white blood cells into groups based on appearance, rather than function, the two groups are granulocytes and agranulocytes. The granulocytes are more granular in appearance compared to the agranulocytes. Granulocytes include neutrophils, eosinophils, and basophils. The agranulocytes include monocytes and lymphocytes. Monocytes are the only phagocytes that are not granulocytes.

4.5.1 Neutrophils

Neutrophils are the most abundant type of leukocyte and are granulocytes. Neutrophils have a characteristic appearance with a two to five lobed nucleus. Figure 4.4 shows a photograph of a neutrophil. The lifespan of neutrophils is around 10 h. Neutrophils are motile and phagocytic and they play a key role in the body's defense against bacterial invasion. They are the first leukocytes to arrive at an area of tissue damage.

Neutrophil leukocytosis is an increase in the number of circulating neutrophils to a level greater than 7500 per mm^3. Neutrophil leukocytosis is the most frequently observed change in the blood count. A few causes of this include inflammation and tissue necrosis, for example, as a result of a myocardial infarction, acute hemorrhage, and bacterial infection. Acute infections such as appendicitis, smallpox, or rheumatic fever can result in leukocytosis. If the neutrophil count becomes considerably less than normal, it can also be due to a viral infection like influenza, hepatitis, or rubella.

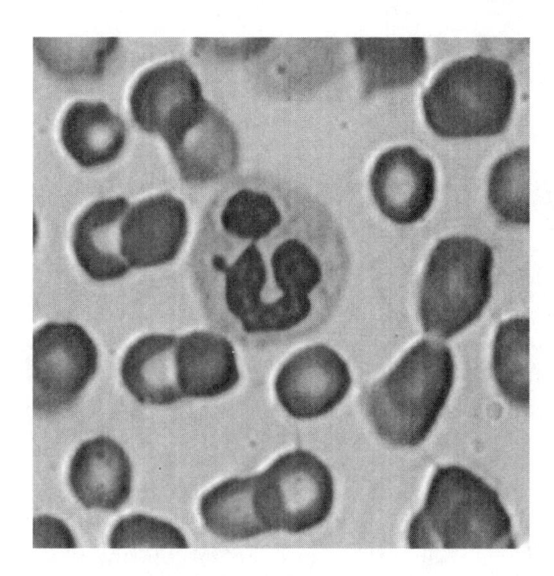

Figure 4.4 A photomicrograph of a neutrophil. Note the four lobed nucleus.

Physiologic leukocytosis occurs in newborn infants, pregnancy and after strenuous exercise, when the number of circulating neutrophils can increase about 7500 per mm^3 with no associated pathology.

4.5.2 Lymphocytes

Lymphocytes are the immunological component of our defense against foreign invasion of the body. The lymphocytes make up about a fourth of the circulating leukocytes. Lymphocytes are nonphagocytic and a primary function of these cells is the release of antibody molecule and antigen disposal.

B lymphocytes originate in the bone marrow and make up about 20 percent of lymphocytes. These cells synthesize antibody molecules. The type of immunity that involves B lymphocytes is known as humoral immunity. The B cells fight against bacteria.

T lymphocytes on the other hand originate in the thymus gland and make up about 80 percent of lymphocytes. These T cells are preconditioned to attack antigens either directly or by releasing chemicals to attract neutrophils and B lymphocytes. T lymphocytes do not produce antibodies directly. On the other hand, T cells multiply and clone more T cells which are also responsive to some antigen. This is known as cell mediated immunity. T cells fight against bacteria, viruses, protozoa, and fungi. T cells also fight against things like transplanted organs. Patients with acquired immune deficiency syndrome (AIDS) monitor their T-cell levels; an indicator of the AIDS virus' activity. Figure 4.5 shows a photomicrograph of a lymphocyte.

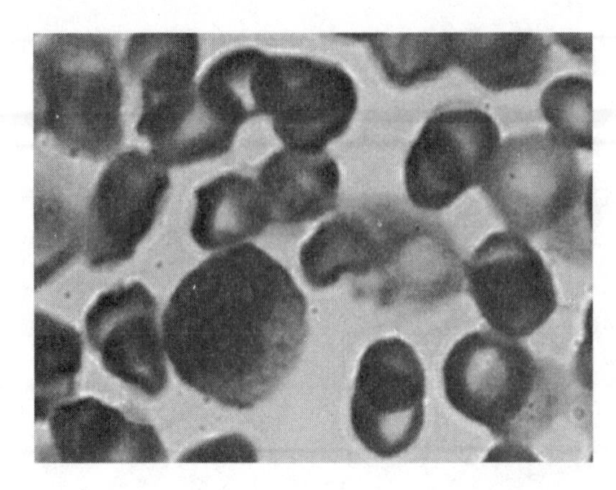

Figure 4.5 A photomicrograph of blood showing a lymphocyte.

Lymphocytosis is defined as an increase in the number of circulating lymphocytes above normal. There are normally 1000 to 3000 lymphocytes in a cubic millimeter of blood. When the number of circulating lymphocytes becomes 10,000 per mm^3 for example, this is known as lymphocytosis. Some causes of acute lymphocytosis include infectious mononucleosis, rubella, mumps, and HIV. Some causes of chronic lymphocytosis might include tuberculosis or syphilis.

The life span of a lymphocyte is weeks to months. A large number of inherited or acquired deficiencies in any of the components of the immune system may cause an impaired immune system response. A primary lack of T lymphocytes, as in AIDS, leads not only to bacterial infections, but also to viral, protozoal, and fungal infections.

4.5.3 Monocytes

Monocytes are mobile and actively phagocytic and are larger than other leukocytes. They have a large central oval or indented nucleus. They circulate in the blood stream for 20 to 40 h and then leave the blood stream to enter the body's tissues and mature and differentiate into macrophages. Macrophages carry out a function similar to that of a neutrophil. Macrophages may live for several months or even years!

Three to nine percent of all circulation leukocytes are monocytes. They increase in number in patients with malaria, mononucleosis, typhoid fever, and Rocky Mountain spotted fever.

4.5.4 Eosinophils

Eosinophils are granulocytes and they have a special role in allergic response or defense against parasites and in the removal of fibrin formed during inflammation. The primary function of eosinophils is detoxification of foreign proteins. The number of eosinophils in whole blood is normally zero to three percent of leukocytes. However, in allergic reactions this percentage increases. Eosinophils increase in numbers due to bronchitis, bronchial asthma, or hay fever.

4.5.5 Basophils

Basophils are only occasionally seen in normal peripheral blood. One function of a basophil is to release histamine in an area of tissue damage in order to increase blood flow and to attract other leukocytes to the area of damage. Basophils increase in number in response to hemolytic anemia or chicken pox.

4.5.6 Leukemia

Leukemia is a group of disorders characterized by the accumulation of abnormal white cells in the bone marrow. Another way to say this is that there is a purposeless malignant proliferation of leukopoietic tissue. Leukopoietic tissue is the tissue, which forms leukocytes. This type of leukopoietic tissue is found in bone marrow, lymph nodes, spleen, and thymus for example. Some common features of leukemia include a raised total white cell count, abnormal white blood cells in peripheral blood, and evidence of bone marrow failure.

4.5.7 Thrombocytes

Thrombocytes are also known as platelets. Platelets come from megakaryocytes which are giant (30 μm) cells from bone marrow. The mean diameter of a platelet is about 1 to 2 μm or about 1/4 to 1/8 the diameter of an erythrocyte.

The normal lifespan of thrombocytes is seven to ten days. The normal platelet count in healthy humans is about 250,000 platelets per cubic millimeter of whole blood. Young platelets spend up to 36 h in the spleen after being released from the bone marrow.

Platelets are granular in appearance and have mitochondria but no nucleus. The main function of platelets is the formation of mechanical plugs during normal hemostatic response to vascular injury. Platelet reactions of adhesion, secretion, aggregation, and fusion are important to this hemostatic function. Thrombocytes are shown in Fig. 4.6 and are marked with a "t."

4.6 Blood Types

ABO blood types are genetically inherited. To understand blood types, we must first learn a few vocabulary words related to the immune system and genetics. Antibodies, or immunoglobulins, are a structurally related

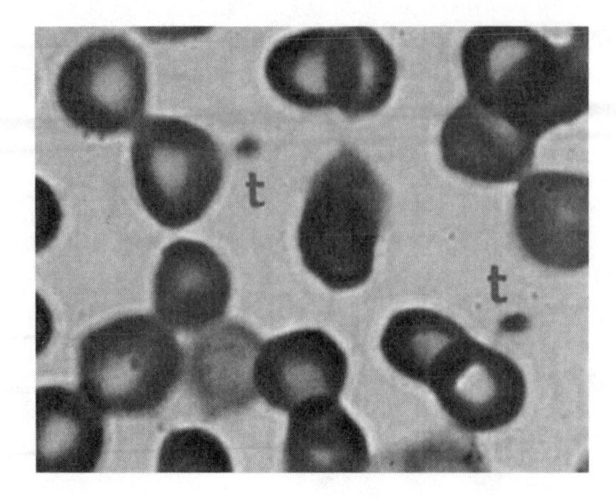

Figure 4.6 Shows a photomicrograph of blood showing thrombocytes, marked "t."

class of glycoprotein, which are produced in response to a specific antigen. Antigens on the other hand, are any substance, which attracts antibodies. This could include foreign proteins, toxins, bacteria, and virus.

An allele is any alternative form of a gene that can occupy a particular chromosomal locus. A genotype is the genetic makeup of an individual; or the alleles present at one or more specific loci. A phenotype is the entire physical, biochemical, and physiological makeup of an individual (as opposed to genotype).

The three alleles associate with the ABO blood groups are I^A, I^B, and i. Alleles I^A and I^B are dominant to i, but show no dominance with respect to one another. Each person has two alleles, one inherited from each parent. For example, if a person inherits the I^A allele from each parent, he or she will have the I^A/I^A genotype and blood type A (phenotype).

Agglutination is the term for the aggregation of erythrocytes into clumps. Severe agglutination can lead to death. Agglutinogens (or antigens) are substances on the membranes of erythrocytes. Agglutinins are antibodies in the plasma, which try to attack specific agglutinogens. This is known as an antibody-antigen reaction.

Persons with type A blood have the I^A/I^A genotype and their erythrocytes have the A antigen (agglutinogen) and the β or anti-B antibody (agglutinin). If this person received a transfusion of type B blood, his/her antibodies would attack the antigens on the donor erythrocytes and cause agglutination.

Persons with type B blood have the B antigen. Persons with type AB blood have both the A and the B antigens on their erythrocytes. Finally, people with type O blood have no A or B antigens on their erythrocytes.

Table 4.2 shows phenotypes and genotypes of the four blood groups along with their respective antigens and antibodies. Note that persons with A blood have the β antibody, persons with type B blood have the α

TABLE 4.2 Blood Types, Distributions and Their Associate Antigens and Antibodies

Blood type phenotype	Blood group genotype	Erythrocyte antigen (agglutinogens)	Plasma antibody (agglutinins)	Population distribution whites, Iowa (for example)
A	I^A/I^A or I^A/i	A	Anti-B (β)	42%
B	I^B/I^B or I^B/i	B	Anti-A (α)	9%
AB	I^A/I^B	A and B	Neither	4%
O	i/i	Neither	Anti-A and Anti-B (α and β)	46%

antibody and persons with O blood have both the α and β antibodies. Figure 4.7 shows the relative population distributions of the various blood types among white North Americans.

Persons with AB blood have no antibodies and are therefore known as universal recipients. Persons with type O blood have both antibodies but no A or B antigens on their red blood cells and are therefore known as universal donors.

H is a weak antigen and nearly all people have it. The I^A and I^B alleles cause H to be converted to the A or B antigen. On the other hand, people with type O have more H antigen since it is not converted.

4.6.1 Rh blood groups

Rh antigen was named for the rhesus monkey in which it was first detected. Today over 40 Rh antigens are known in humans. One of these antigens is a strong antigen and can lead to transfusion problems.

A person who is Rh negative does not have the Rh antigen and normally does not have the antibody. Upon exposure to the Rh antigen, Rh negative persons develop the antibody. An Rh− mother can produce antibodies after the birth of an Rh+ neonate after mixing of the mother's

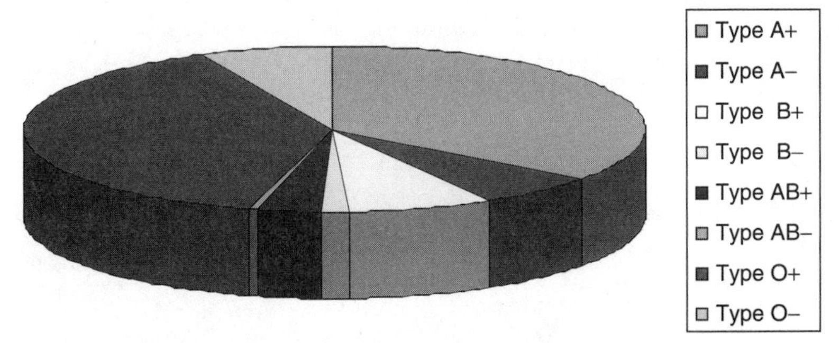

Figure 4.7 Shows the population distribution of ABO blood types among white Iowans.

blood and the newborn's blood occurs. If that Rh−ve mother with antibodies has a second pregnancy and the fetus is Rh+ve, the antibodies will cross over the placenta and begin to destroy fetal erythrocytes (erythroblastosis fetalis). Some of these babies can be saved by transfusion.

Tests can now reveal the presence of the fetal Rh+ve cells in the blood stream of the Rh−ve mother. When such cells are present, the expectant mother is given injections of the antibody. Because the antibody reacts with and covers up the antigen, the mother's immune system is not stimulated to produce its own antibodies.

4.6.2 M and N blood group system

Many other antigens exist and although they may not be medically important in transfusions, they can be used to study genetics and are used in tests involving disputed parenthood.

Two antigens, M and N exist and the alleles for production of these antigens show no dominance to each other. There are therefore three corresponding genotypes: M, MN, and N. The M phenotype is the blood type with the M antigen located on the erythrocyte and is associated with the M/M genotype. Table 4.3 shows a list of MN phenotypes along with their respective genotypes and antigens.

4.7 Plasma

Plasma is a transparent amber fluid and is 90 percent water by volume. The 10 percent that is not water on the other hand adds some very important characteristics to the function of blood.

Plasma contains inorganic substances like sodium ions, potassium ions, chloride ions, bicarbonate ions, calcium ions and those chemicals make up about 1 percent of the plasma by volume. Plasma proteins make up about 7 percent of plasma by volume. Those proteins include albumins, globulins, and fibrinogens. In plasma there is a further ~1 percent by volume of nonprotein organics and also varying amounts of hormones, enzymes, vitamins, and dissolved gases.

Plasma proteins are large molecules with high molecular weight that do not pass through the capillary wall. They remain in the blood vessel and establish an osmotic gradient.

TABLE 4.3 MN Phenotypes and Their Associated Genotypes and Antigens

Phenotype	Genotype	Erythrocyte antigen
M	M/M	M
MN	M/N	M and N
N	N/N	N

Albumin accounts for most of the plasma protein osmotic pressure. Human plasma contains about 4 to 5 g of albumin per 100 mL of plasma. Albumin is important in binding certain substances that are transported in plasma, such as barbiturates and bilirubin.

Globulins are present in human plasma at the rate of approximately 2 to 3 g per 100 mL plasma. The normal albumin/globulin ratio is approximately two. Fibrinogen is a plasma protein that is involved in hemostasis. Hemostasis is the process by which loss of blood from the vascular system is reduced (blood clotting).

4.7.1 Plasma viscosity

Plasma viscosity is a function of the concentration of plasma proteins of large molecular size. This is particularly true of the proteins with pronounced axial asymmetry. Fibrinogen and some of the immunoglobulins fall in this category. Normal values of viscosity at room temperature are in the range of 1.5 to 1.7 cP (1.5 to 1.7×10^{-3} Ns/m^2).

4.7.2 Electrolyte composition of plasma

One equivalent (eq) contains Avogadro's number of positive or negative charges (6×10^{23} charges).

Therefore, for singly charged (univalent ions) one equivalent of ions is equal to 1 mole and therefore 1 meq = 1 mmol.

Table 4.4 shows the electrolyte composition of plasma. Note that sodium is an important ion in plasma, which contains 142 meq/L. This is the same as 142 mmol/L. Plasma also contains 5 meq/L of calcium ion, which is the same as 2.5 mmol per liter because calcium is a doubly charged ion.

Osmotic pressure is defined as the pressure that builds up as a result of the tendency of water to diffuse down the concentration gradient. On the other hand, oncotic pressure is osmotic pressure due to plasma proteins.

Osmolarity, in units of Osm/L, is the measure of the concentration of a solute, which would cause an osmotic pressure. Osmolarity is equal to molar concentration multiplied by the number of ionized particles in

TABLE 4.4 Electrolyte Composition of Plasma

Substance	Symbol	meq/L	Substance	Symbol	meq/L
Sodium	Na^+	142	Chlorine	Cl^-	103
Potassium	K^+	4	Bicarbonate	HCO_3^-	28
Calcium	Ca^{++}	5	Proteins		17
Magnesium	Mg^+	2	Others		5
Total		153			153

the solution. When a solute has the concentration of 1 Osmole (Osm) it would have an osmotic pressure of 22.4 atm compared to pure water. Human plasma has the same osmotic pressure as 0.9 percent NaCl solution (physiologic or isotonic saline).

0.9 percent NaCl is a solution that has 9 g of NaCl in 1000 g of water or 9 g of NaCl per liter of water. One mole of NaCl contains 23 g of solium and 35 g of chlorine for a total of 58 g of sodium chloride. The molar concentration of isotonic saline is 9 g of sodium chloride in 1 L of water or 9/58 = 155 mmolar. The osmolarity of isotonic saline is 155 mmolar multiplied by 2 ionized particles (sodium and chloride) or 310 mOsm.

When erythrocytes are placed in an isotonic solution, they do not change size. When they are placed in a hypertonic solution, water passes out of the cell and they shrink. When the erythrocytes are placed in a hypotonic solution (0.35 percent NaCl for example) they swell to a nearly spherical shape and in some cases undergo hemolysis.

4.8 Blood pH

The pH or normal healthy blood is in the range of 7.35 to 7.45. When the pH is less than 7.35 this condition is defined as acidosis. When the pH is greater than 7.45, this condition is defined as alkalosis. CO_2 dissolved in water in plasma produces carbonic acid, which lowers blood pH. Bicarbonate and carbonic acid form an acid–base buffer pair, which help to keep the arterial pH near 7.4. When PCO_2 decreases, then the pH increases.

The ratio of bicarbonate concentration to partial pressure of CO_2 in blood is a ratio of metabolic compensation to respiratory compensation. The Henderson Hasselback equation can be used to calculate the pH of arterial blood based on that ratio.

$$\mathrm{pH} = 6.1 + \log \frac{[\mathrm{HCO_3^-}]}{0.03(\mathrm{PCO_2})} \tag{4.4}$$

Where the bicarbonate concentration is given in milliequivalents per liter and the partial pressure of carbon dioxide is given in millimeters of mercury. The normal ratio of metabolic compensation to respiratory compensation can be estimated as follows:

$$\text{Normal ratio} = \frac{(\mathrm{HCO_3^-})}{0.03(\mathrm{PCO_2})} = \frac{24}{0.03(40)} = 20$$

4.9 Clinical Features

An imbalance of acid generation or removal due to either lungs or kidneys is usually compensated by a nearly equal and opposite change in the other organ. For example, diarrhea with a loss of bicarbonate-containing fluids will trigger hyperventilation and respiratory compensation to eliminate enough CO_2 to compensate for the bicarbonate loss in the stool. Conversely, vomiting with loss of stomach hydrochloric acid results in metabolic alkalosis for which hypoventilation and an increase in blood CO_2 is the compensation.

Bibliography

Hoffbrand AV, Pettit JE., *Essential Haematology*. 3rd ed. Blackwell Science; 1998.
Selkurt EE. *Basic Physiology for the Health Sciences*. Boston, Little, Brown & Co; 1982.

Chapter

5

Anatomy and Physiology of Blood Vessels

5.1 Introduction

Arteries are the high-pressure blood vessels that transport blood from the heart through increasingly smaller arteries, to arterioles and further to the level of capillaries. Veins conduct the blood from the capillaries back to the heart on the lower pressure side of the cardiovascular system. The structure of arteries and veins as well as their mechanical properties are discussed in this chapter. At any given time, about 13 percent of the total blood volume resides in the arteries and about 7 percent resides in the capillaries.

The structure of veins, arteries, and capillaries differ because each are specialized to perform their respective perfusion, exchange, and capacitance function. However, the inner layer of all blood vessels is lined with a single layer of endothelial cells.

5.2 General Structure of Arteries

There are three types of arteries that can be classified according to their structure and size. Elastic arteries, which include the aorta, have a relatively greater diameter and more elastic fibers. Muscular arteries are smaller in diameter than elastic arteries but larger than arterioles and have a relatively larger proportion of muscle compared to connective tissue. Arterioles are the smallest diameter arteries and have a few layers of smooth muscle tissue and almost no connective tissue.

In general arteries are composed of three layers—the tunica intima or the innermost layer, the tunica media or middle layer, and the tunica

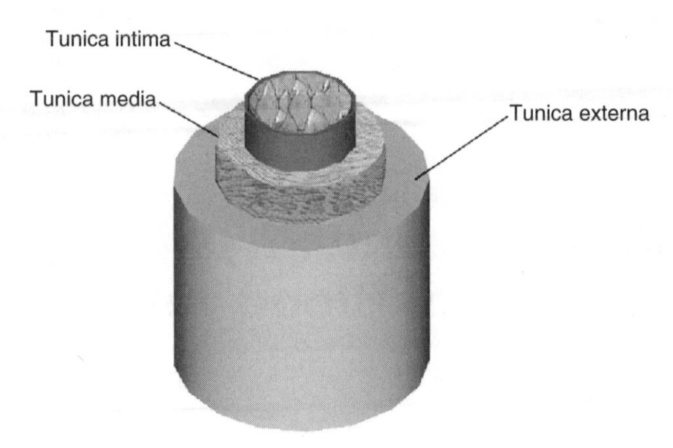

Figure 5.1 Structure of an artery.

externa, which is the outermost layer of the artery. The three layers of an artery can be seen in Fig. 5.1.

The lumen of an artery is the inside of the vessel where blood flows. The lumen is lined with the endothelium that forms the interface between the blood and the vessel wall. The endothelium consists of simple squamous epithelium cells that line the lumen. The cells are in close contact and form a slick layer that prevents blood cell interaction with the vessel wall as blood moves through the vessel lumen.

5.2.1 Tunica intima

The endothelium plays an important role in the mechanics of blood flow, blood clotting, and leukocyte adhesion. For years, the endothelium was thought of as an inert single layer of cells that passively allowed the passage of water and other small molecules across the vessel wall. Today it is clear that it performs many other functions like the secretion of vasoactive substances and the contraction and relaxation of vascular smooth muscle. The innermost layer of an artery that is composed of endothelium is known as the tunica intima. The tunica intima is one cell layer thick and it is composed of endothelial cells.

5.2.2 Tunica media

The tunica media or "middle coat" is the middle layer of a blood vessel. The tunica media consists primarily of smooth muscle cells. The tunica media is living and active. The tunica media can contract or expand and change diameters of the vessel, allowing a change in blood flow. Vasoconstriction is the word used to describe a reduction in vessel diameter due to muscular contraction. At a given pressure, the diameter of

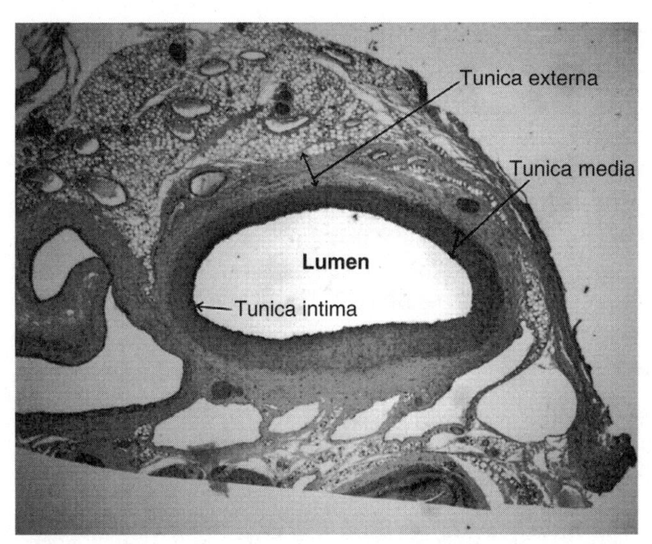

Figure 5.2 Drawing of a photomicrograph of the cross section of an artery showing the tunica intima, tunica media, and tunica externa.

the vessel increases when smooth muscle in the tunica media relaxes. This increase allows more blood flow for the same driving pressure and this process is known as vasodilation.

The tunica media also consists of elastic tissue. This tissue is passive, does not consist of living cells, and does not have significant metabolic activity. The elastic fibers that make up the tunica media support the blood vessel but also allow recoil of an expanded blood vessel when the pressure is removed. The tunica media is absent in capillaries.

5.2.3 Tunica externa

The tunica externa is also sometimes called the tunica adventitia. The tunica externa is composed of connective tissue including passive elastic fibers like in the tunica medica. However, the tunica externa also contains passive, much stiffer, collagenous fibers. Just like the tunica media, the tunica externa is absent in capillaries. Figure 5.2 shows a photomicrograph of the cross section of an artery.

5.3 Types of Arteries

Although arteries have the same general structure, they can be divided into groups by their specific functional characteristics. The three types of arteries covered in this chapter are elastic arteries, muscular arteries, and arterioles.

5.3.1 Elastic arteries

Elastic arteries have the largest diameter of the three groups of arteries. The aorta is a good example of an elastic artery. One of the chief characteristics of elastic arteries is their ability to stretch and hold additional volume, thus performing the function of a "capacitance" vessel. Just as an electrical capacitor stores charge when given an increased voltage, an elastic vessel can store additional volume when subjected to an increased pressure. Elastic arteries have very thick tunica medias when compared with other arteries and contain a lot of the elastic fiber, elastin.

5.3.2 Muscular arteries

Muscular arteries are intermediate-sized arteries. In these vessels the tunica media is composed almost entirely of smooth muscle. The tunica media of a muscular artery can be up to 40 cell layers thick. Functionally, muscular arteries can change diameter to influence flow through vasoconstriction and vasodilation. Most arteries are muscular arteries.

5.3.3 Arterioles

Arterioles are defined as those arteries that have a diameter of less than 0.5 mm. Arterioles have a muscular tunica media that is one to five layers thick. These layers are composed entirely of smooth muscle cells. Arterioles also do not possess much of a tunica externa.

5.4 Mechanics of Arterial Walls

To understand the mechanics of arterial walls, begin by imagining a long tube of constant cross section and constant wall thickness. This imaginary tube is homogenous and isotropic (the material properties are identical in all directions). The typical blood vessel is branched and tapered. It is also nonhomogeneous and nonisotropic. Although our assumptions do not fit the blood vessel strictly speaking, it makes a practical first estimate of a model to help understand vessel mechanics. We need to continually remember our assumptions and the limitations they bring to the model as we progress in our knowledge of blood vessel mechanics.

Consider a cross section of an artery as shown in Fig. 5.3 with wall thickness h, and inside radius r_i. Blood vessels are borderline thin-walled pressure vessels. Thin-walled pressure vessels are those with a thickness to radius ratio that is less than or equal to about 0.1. For arteries, the ratio of wall thickness h to inside radius r_i is typically between 0.1 and 0.15, as shown in Eq. (5.1).

$$\frac{h}{r_i} \cong 0.1 \text{ to } 0.15 \tag{5.1}$$

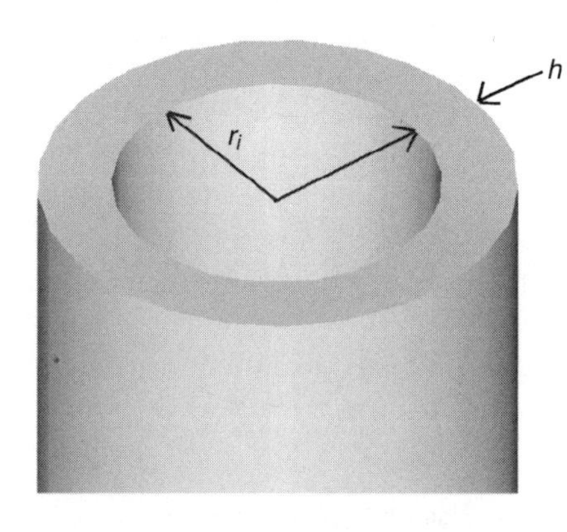

Figure 5.3 An artery modeled as a homogenous long, straight tube with constant cross section and constant wall thickness.

The mechanics of the artery are not dependent only on geometry, but rather to understand the mechanics of the artery we must also consider material properties. In order to consider material properties, let us begin with Hooke's law for uniaxial loaded members. Hooke's law for this simple loading condition relates stress to strain in a tensile specimen. Hooke's law for a one-dimensional uniaxial loaded member is shown below in Eq. (5.2).

$$\sigma = E\varepsilon \tag{5.2}$$

where σ is the normal stress in N/m^2, E is the modulus of elasticity, N/m^2, and ε is the strain, which is unitless.

Figure 5.4 shows a stress-strain curve for a typical engineering material. The modulus of elasticity is the slope of the stress strain curve. For

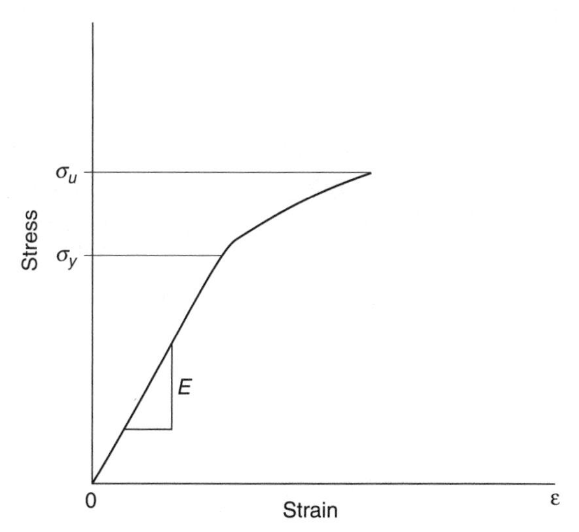

Figure 5.4 Example stress-strain curve for a linearly elastic material.

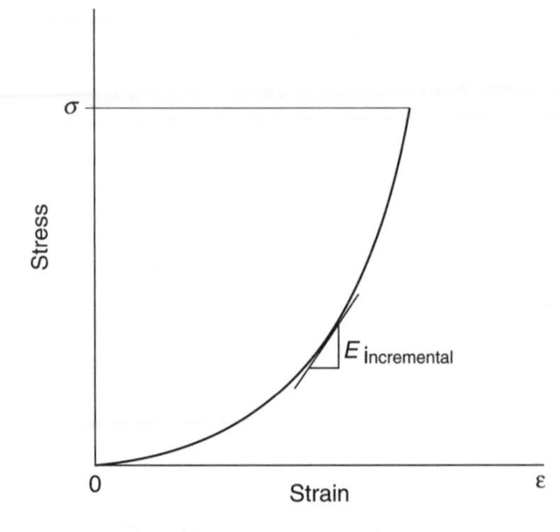

Figure 5.5 Stress-strain curve for an artery.

linearly elastic materials, the modulus of elasticity, E, is a constant in the linearly elastic range.

Arteries are not linearly elastic. The stress-strain curve for an artery is shown in Fig. 5.5. It is still possible to define the modulus of elasticity, but the slope of the curve varies with stress and strain. As stress increases in an artery the material becomes stiffer and resists strain. Arteries are also metabolically active materials. Smooth muscle can contract and expend energy in an effort to resist strain.

Now that we are considering a material in which the modulus of elasticity is not a constant, Hooke's law no longer applies. Instead, the modified Hooke's law, as shown in Eq. (5.3), describes the material behavior:

$$\sigma = \int_0^{E_f} E_{\text{inc}} d\varepsilon \tag{5.3}$$

Also, arteries are viscoelastic materials. Viscoelasticity is a material property in which the stress is not only dependent on load and area, but also on the rate of strain. For a material in which the stress is dependent on the rate of strain, Eq. (5.4) is true:

$$\sigma = E_1 \varepsilon + E_2 \frac{d\varepsilon}{dt} \tag{5.4}$$

5.5 Compliance

To understand the material property that relates a vessel diameter to the pressure inside that tube, begin by considering a section of the tube shown in Fig. 5.6. In the figure the cross section of a tube is shown where r_i is the inside radius of the artery, r_o is the outside radius of the artery,

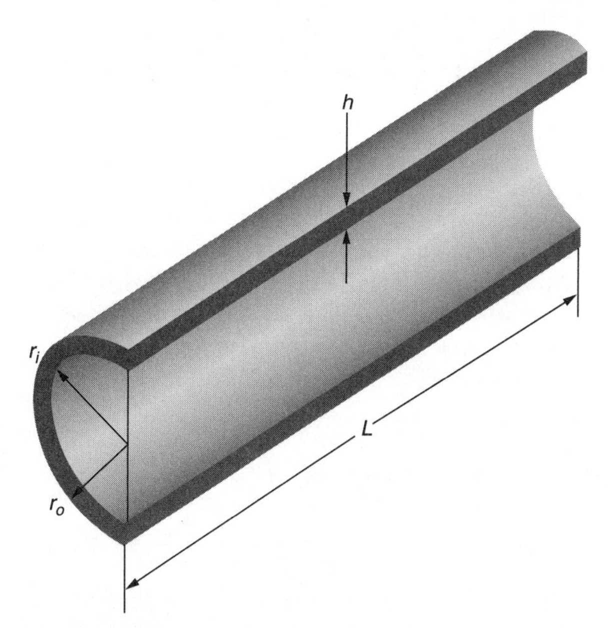

Figure 5.6 Isometric drawing of a tube used to model compliance.

and h is the thickness of the tube, $h = r_o - r_i$. Figure 5.7 shows a free-body diagram of the same tube on which we will now perform a force balance.

$$\sum F_x = 0$$

$$P_o 2 r_o L + 2 S_h (r_o - r_i) L = P_i 2 r_i L \tag{5.5}$$

$$S_h (r_o - r_i) = (P_i r_i) - (P_o r_o) \tag{5.6}$$

$$S_h h = (P_i r_i) - (P_o r_o) \tag{5.7}$$

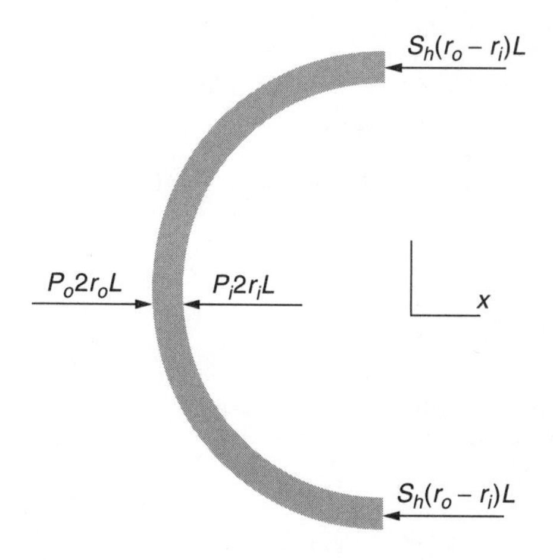

Figure 5.7 Cross-sectional drawing of a tube used to model compliance.

We now define wall tension, T, which is equivalent to the hoop stress in a pressure vessel multiplied by the wall thickness. Wall tension has units of force per length.

$$S_h h \equiv T \text{ (force per unit length)} \tag{5.8}$$

From mechanics of materials, we know something about the relationship between stress, pressure, and tube geometry. Replacing $S_h h$ with T yields the law of Laplace as written in Eq. (5.9). Let us also call $P_i - P_o$ simply transmural pressure, P. For thin walled pressure vessels, where the outside diameter is approximately equal to the inside diameter and where the ratio of radius to wall thickness is greater than 10, then Eq. (5.10) applies.

$$T = Pr \quad \text{Law of Laplace} \tag{5.9}$$

$$S_h = \frac{Pr}{h} \tag{5.10}$$

As the pressure inside the vessel increases, the radius increases, as shown in Fig. 5.8. The resulting strain, $d\varepsilon$, is equal to the change in radius divided by the radius, as shown in Eq. (5.11).

$$d\varepsilon = \frac{dr}{r} \tag{5.11}$$

Also, the modulus of elasticity, E, for thin walled pressure vessels is equal to the ratio of the hoop stress, S_h, divided by the strain, as shown in Eq. (5.12).

$$E = \frac{dS_h}{d\varepsilon} \tag{5.12}$$

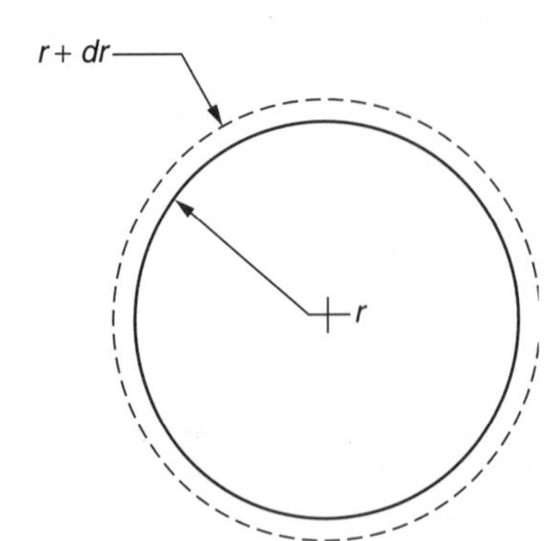

Figure 5.8 Change in vessel radius as a result of increasing pressure.

Equations (5.13) and (5.14) relate the initial and final radii and wall thicknesses. The equation is based on conservation of mass. Equation (5.14) assumes constant density for the blood vessel material since blood vessels are essentially incompressible. Equation (5.14) also assumes that the length of the blood vessel does not change significantly.

$$2\pi r_o h_o l_o \rho_o = 2\pi r h l \rho \tag{5.13}$$

$$r_o h_o = rh \tag{5.14}$$

In both equations r_o is the initial vessel radius and h_o is the initial vessel wall thickness. In both equations, r and h are the vessel radius and wall thickness at any other point in time. Combining Eqs. (5.11) and (5.12) we can now write

$$dS_h = E\frac{dr}{r} \tag{5.15}$$

By combining Eqs. (5.10) and (5.14) it is also possible to write

$$dS_h = d\left(\frac{Pr^2}{r_o h_o}\right) \tag{5.16}$$

$$d(Pr^2) = 2Prdr + r^2 dP \tag{5.17}$$

Since for small strains $dr \sim 0$, we can combine Eqs. (5.15), (5.16), and (5.17) into a single equation and solve for the pressure change as a function of radius:

$$E\frac{dr}{r} = \frac{r^2 dP}{r_o h_o} \tag{5.18}$$

$$\int_{P_o}^{P} dP = Eh_o r_o \int_{r_o}^{r} \frac{dr}{r^3} \tag{5.19}$$

$$P - P_o = \frac{Eh_o r_o}{2}\left[\frac{1}{r_o^2} - \frac{1}{r^2}\right] \tag{5.20}$$

$$P - P_o = \frac{Eh_o r_o}{2}\left[1 - \frac{A_o}{A}\right] \tag{5.21}$$

In Eq. (5.21) A_o is the initial cross-sectional area of the vessel and A is the area of the vessel at time equals t. In Eq. (5.22), we then solve for the area ratio.

$$\frac{A}{A_o} = \left[1 - \frac{(P - P_o)2r_o}{Eh_o}\right]^{-1} \tag{5.22}$$

It is also possible to simplify this expression as a polynomial series since the expression $(P - P_o)2r_o/Eh_o < 1$ for practical, physiological circumstances. The form of the polynomial series is

$$[1 - x]^{-1} = 1 + x + x^2 + x^3 + \cdots$$

Let $2r_o/Eh_o = C_1$

$$\frac{A}{A_o} = 1 + C_1(P - P_o) + [C_1(P - P_o)]^2 + [C_1(P - P_o)]^3 + \cdots \quad (5.23)$$

Taking the derivative of the cross-sectional area of the vessel with respect to pressure, yields Eq. (5.24):

$$\frac{dA}{dP} = A_oC_1 + A_oC_1^2\, 2(P - P_o) + A_oC_1^3\, 3(P - P_o)^2 + \cdots \quad (5.24)$$

Practically speaking, $C_1^2, C_1^3, \ldots$ are so small that we can now estimate the compliance of the vessel by Eq. (5.25).

$$C = \frac{dA}{dP} \approx A_oC_1 = \frac{2\pi r_o^3}{Eh_o} \quad (5.25)$$

Finally, it is also possible to estimate modulus of elasticity if the compliance is known. In Eq. (5.26), modulus of elasticity is derived from Eq. (5.25).

$$E = \frac{2\pi r_o^3}{Ch_o} \quad (5.26)$$

Example In Fig. 5.9 a tube with modulus of elasticity, E, is anchored between two stationary ports and injected with a known volume of water. For the pressure increase $(P_f - P_o)$, fixed length L_o, tube diameter D, and volume of injected fluid V, calculate the compliance and estimate the modulus of elasticity. The wall thickness of the tube is 1 mm.

- Initial pressure, $P_o = 10$ kPa
- Pressure increase, $P_f - P_o = 6$ kPa
- Initial tube diameter, $D = 20$ mm
- Initial length, $L_o = 25$ cm
- Volume injected, $V = 10$ mL

The change in cross-sectional area resulting from an injection of 10 mL into the tube is the change in volume divided by the fixed tube length.

$$\Delta A = \frac{\Delta V}{L_o} = \frac{10 \text{ cm}^3}{25 \text{ cm}} = 0.4 \text{ cm}^2$$

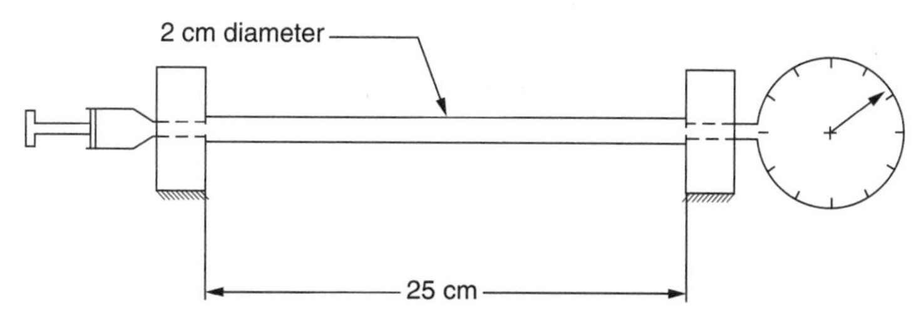

Figure 5.9 Shows a fixture for measuring compliance and estimating modulus of elasticity in a 2 cm diameter tube.

The compliance of the tube can then be calculated as the change in area divided by the change in pressure.

$$\text{compliance} = \frac{dA}{dP} = \frac{0.4\ \text{cm}^2}{6000\ \text{Pa}} \times \frac{1\ \text{m}^2}{(100)^2\ \text{cm}^2} = 6.7 \times 10^{-9}\,\frac{\text{m}^2}{\text{Pa}}$$

Once the compliance has been determined, the incremental modulus of elasticity for the tube can be estimated from the compliance.

$$\text{modulus of elasticity} = E = \frac{2\pi r_o^3}{Ch_o} = \frac{2\pi(0.01)^3\ \text{m}^3}{6.7 \times 10^{-9}\,\frac{\text{m}^2}{\text{Pa}} \times 0.001\ \text{m}}$$

$$= 0.94\ \text{MPa}$$

5.6 Pressure-Strain Modulus

It is difficult to measure the thickness of arteries in vivo, so sometimes scientists also report the pressure-strain modulus. The pressure strain modulus is defined by the ratio of pressure change to normalized diameter change resulting from that pressure change. Equations (5.27) through (5.30) designate the pressure strain modulus E_p.

$$Ep \equiv \frac{P_{\text{max}} - P_{\text{min}}}{(d_{\text{max}} - d_{\text{min}})/d_{\text{mean}}} \tag{5.27}$$

$$E = \frac{2\pi r_o^3 2}{h_o \pi d_{\text{m}}}\left(\frac{P_{\text{max}} - P_{\text{min}}}{d_{\text{max}} - d_{\text{min}}}\right) \tag{5.28}$$

$$E = \frac{2\pi r_o^2}{r_o}\left(\frac{P_{\text{max}} - P_{\text{min}}}{d_{\text{max}} - d_{\text{min}}}\right) \tag{5.29}$$

$$E = \frac{r_o}{h_o} E_p \tag{5.30}$$

In 1972 in their paper titled, "Transcutaneous Measurement of the Elastic Properties of the Human Femoral Artery," Mozersky et al.

TABLE 5.1 Pressure-Strain Modulus of Arteries in Humans

Under 35 years	2.64×10^5 N/m^2
35–60 years	3.88×10^5 N/m^2
Over 60 years	6.28×10^5 N/m^2

reported the pressure-strain modulus in humans and how it varied with age. From the Table 5.1 it is possible to see how the stiffness of the blood vessels increased with age.

5.7 Vascular Pathologies

5.7.1 Atherosclerosis

Atherosclerosis is a vascular pathology that has become a prominent disease in western society. The term comes from the Greek words athero (gruel or paste) and sclerosis (hardness) and is characterized by the progressive narrowing and occlusion of blood vessels. When fatty substances, cholesterol, cellular waste products, calcium, and fibrin build up in the inner lining of an artery, this causes a narrowing of the lumen of the vessel and also an increase in the wall stiffness or decrease in compliance of the vessel. The buildup that results is called plaque. Some level of stiffening of the arteries and narrowing is a normal result of aging. Eventually the plaque can block an artery and restrict flow through that vessel, resulting in a heart attack if the vessel being blocked is one that supplied blood to the heart. Atherosclerosis can also produce blood clot formation, sometimes resulting in a stroke. When a piece of plaque breaks away from the arterial wall and flows downstream, it can also become lodged in smaller vessels and block flow, also resulting in stroke.

Atherosclerosis usually affects medium sized or large arteries. Three risk factors associated with atherosclerosis are elevated levels of cholesterol and triglycerides in the blood, high blood pressure, and cigarette smoking.

5.7.2 Stenosis

Stenosis refers to an obstruction of flow through a vessel. When a localized plaque forms inside a vessel, this is called a stenosis. An aortic stenosis, for example, refers to an obstruction at the level of the aortic valve. An aortic stenosis is typically seen as a restricted systolic opening of the valve with an increased pressure drop across the valve. Doppler echocardiography can be used to identify and quantify the severity of a valvular stenosis.

5.7.3 Aneurysm

Aneurysm is the term that refers to the abnormal enlargement or bulging of an artery wall. This condition is the result of a weakness or thinning of the blood vessel wall. Aneurysms can occur in any type of blood vessels, but they usually occur in arteries. Aneurysms commonly occur in the abdomen or the brain. These are known as abdominal aortic aneurysms (AAA) or cerebral aneurysms (CA).

An AAA is the enlargement of the lower part of the aorta. AAAs are known as silent killers because in most cases there are no symptoms associated with the pathology. When an AAA ruptures, it will result in life-threatening blood loss. Most AAAs are diagnosed during a physical exam when the person is undergoing a check-up for some other health concern.

AAAs that are less than 2 in. in diameter are usually monitored and treated with blood pressure-lowering drugs. AAAs that are greater than 2 in. in diameter are typically surgically replaced with a flexible graft, when they are discovered.

A cerebral aneurysm is a weak, bulging spot in an artery of the brain. These often result from a weakness in the tunica media (muscle layer) of the vessel that is present from birth. A cerebral aneurysm may cause symptoms ranging from headaches, drowsiness, neck stiffness, nausea, and vomiting to more severe symptoms such as mental confusion, vertigo (dizziness), and loss of consciousness. A ruptured aneurysm most often results in a severe headache demanding immediate medical attention. A cerebral aneurysm can be diagnosed by imaging tests such as x-rays, ultrasound, computed axial tomography (CAT) scans, and magnetic resonance images (MRIs). When detected, brain aneurysms can be treated by microsurgery.

5.7.4 Thrombosis

Thrombosis refers to the formation or development of an aggregation of blood substances, including platelets, fibrin, and cellular elements. Simply put, a thrombus is a blood clot. Thrombosis often results in vascular obstruction at the point of formation. For comparison, an embolism is a general clot or plug that could refer to a blood clot, or an air bubble, or even a mass of cancer cells, that is brought by flowing blood from one vessel into a smaller diameter vessel, where it is eventually lodged.

5.7.5 Clinical aspects

In an article published by Herrington et al. (2004), "Relationship between arterial stiffness and subclinical aortic atherosclerosis," in the journal *Circulation*, noninvasive measures of arterial compliance were made in 267 subjects and compared to the extent of atherosclerosis of the aorta. They found the average of calf and thigh compliance to be strongly predictive of the extent of atherosclerosis. This type of noninvasive measurement of compliance can prove useful in predicting which patients will develop atherosclerosis, while their symptoms are still subclinical.

5.8 Stents

A stent is a metal mesh tube that is inserted into an artery on a balloon catheter and inflated to expand and hold open an artery, so that blood can flow more easily through the artery. The stent remains permanently in place to hold the artery open. Drug eluting stents are stents that contain drugs that are eluted into the blood stream and help prevent the arteries from becoming reclogged.

Stents often become necessary when an artery has narrowed due to atherosclerosis. During this vascular disease, plaque builds up on the endothelium. When the vessel becomes 85 to 90 percent blocked, blood flow becomes restricted and the patient may experience symptoms indicating a decrease in blood flow.

Typically, stents are inserted on a balloon catheter through the femoral artery or brachial artery and guided up to the narrowed section of a smaller artery. The balloon catheter allows the stent to be expanded into place, by the inflation of the balloon after the stent has reached the correct location.

5.9 Coronary Artery Bypass Grafting

Coronary arteries are the blood vessels that deliver blood to heart muscle tissue. When atherosclerotic plaque builds up on the wall of the coronary arteries then the result is coronary artery disease. Plaque accumulations can be accelerated by smoking, high blood pressure, elevated cholesterol, and diabetes. In some cases, patients with coronary artery disease can be treated with a minimally invasive procedure called an angioplasty. During an angioplasty a stent is placed inside the vessel to increase the lumen diameter and to hold the vessel open. When that procedure is not possible, the patient may be a candidate for a coronary artery bypass graft (CABG) surgery. CABG surgery is performed about 350,000 times each year in the United States. When a patient undergoes a bypass graft, the surgeon makes an incision in the middle of the chest and saws through the sternum. The heart is cooled with iced saline while a solution is injected into the arteries to minimize damage.

Veins are typically harvested from the legs to be used as the bypass vessels. It is also possible to use chest arteries, like the internal mammary artery. During the surgery, the bypass vessels are sewed onto the coronary arteries beyond the narrowing or blockage. The other end of the bypass vessel is then attached to the aorta to provide a strong blood supply to the coronary circulation that is unimpeded by blockages.

The entire CABG procedure takes about 4 h to complete. During those 4 h the aorta is clamped for about 60 min and the blood supply to the body is furnished by cardiopulmonary bypass for about 90 min. At the end of surgery, the sternum is wired together with stainless steel wire and the chest incision is sutured closed.

5.9.1 Arterial grafts

Currently there are about 700,000 coronary and 70,000 peripheral arterial grafts each year in the United States. Autologous grafts are grafts that are harvested from a patient and grafted to an artery of the same patient. Typical autologous vessels include the greater saphenous vein and the internal mammary artery.

In 1912, Carrel won the Nobel Prize in medicine for his work on the vascular suture and the transplantation of blood vessels. Then in 1948, Kunlin developed the heart bypass using autologous veins.

Arthur Voorhees (1952) at Columbia University postulated that diseased arteries might be replaced by synthetic fabric. Voorhees and colleagues developed Vinyon-N cloth tubes to substitute for diseased arterial segments. The concept was confirmed and soon synthetic vessels that were larger than 5 mm in diameter were used and often remained patent (unoccluded) for more than 10 years. However, for vessels less than 5 mm in diameter, the prosthetic vessels occlude rapidly upon implantation.

The next advance in vascular grafts came in 1957 when DeBakey and colleagues introduced polyethylene-terephthalate (PET) also known as Dacron. In 1959, W. S. Edwards introduced the polytetrafluoroethylene (PTFE) graft, also known as Teflon. Four years later, in 1969 R. W. Gore invented the expanded polytetrafluoroethylene graft (ePTFE) also known as Gore-Tex.

The search for a viable small diameter vascular graft that is less than 5 mm in diameter continues. By 1978 Herring had described endothelial seeding of synthetic grafts and by 1987, Cryolife Inc., began recovering human greater saphenous veins and cryopreserving them for use as vascular grafts. Grafts in which blood vessels come from a human donor and are matched by blood type between the donor and recipient are known as allografts. In 1988, clinical trials of endothelial sodding on ePTFE grafts took place. Sodding is a high density endothelial seeding in which more than 0.2 million cells per square centimeter are seeded.

One exciting direction for the future of vascular grafts is the area of tissue-engineered vessels. L'Heureux et al. used separately cultured smooth muscle cells and fibroblasts in 1998 to construct a living blood vessel. The smooth muscle was wrapped around a tube to produce the tunica media. A fibroblast sheet was wrapped around that construct to produce the tunica adventitia. After maturation, the tube was removed and endothelial cells were seeded onto the lumen. Viability was demonstrated.

Bibliography

Mozersky DJ, Sumner DS, Hokanson DE, Strandness DE. Transcutaneous Measurement of the Elastic Properties of the Human Femoral Artery, *Circulation*. 1972;46:9

Stoyioglou A, Jaff MR. Medical treatment of peripheral arterial disease: a comprehensive review, *J Vasc Interv Radiol*. 2004;15(11):1197-207.

Herrington DM, Brown WV, Mosca L, Davis W, Eggleston B, Hundley G, Raines J. Relationship between arterial stiffness and subclinical aortic atherosclerosis, *Circulation*. 2004;110:432-437.

Voorhees AB Jr, Janetzky A III, Blakemore AH. The use of tubes constructed from Vinyon "N" cloth in bridging defects, *Ann Surg*. 1952; 135:332.

Selkurt EE. *Basic Physiology for the Health Sciences*. Boston: Little, Brown & Co; 1982.

Chapter

6

Mechanics of Heart Valves

6.1 Introduction

Four cardiac valves help to direct flow through the heart. Heart valves cause blood to flow only in the desired direction. If a heart without heart valves were to contract, it would simply squeeze the blood causing it to flow both backward and forward (upstream and downstream). Instead, under normal physiological conditions, heart valves act as check valves to prevent blood from flowing in the reverse direction. Also, heart valves remain closed until the pressure behind the valve is large enough to cause blood to move forward.

Each human heart has two atrioventricular valves, which are located between the atria and the ventricles. The tricuspid valve is the valve between the right atrium and the right ventricle. The mitral valve is the valve between the left atrium and the left ventricle. The mitral valve prevents blood from flowing backward into the pulmonary veins and therefore into the lungs, even when the pressure in the left ventricle is very high. The mitral valve is a bicuspid valve having two cusps and the tricuspid valve has three cusps.

The atrioventricular valves in the human heart are known as semilunar valves. The two semilunar valves are the aortic valve and the pulmonic valve. The aortic valve is located between the aorta and the left ventricle and when it closes it prevents blood from flowing backward from the aorta into the left ventricle. An aortic valve is shown in Fig. 6.1. The pulmonic valve is located between the right ventricle and the pulmonary artery and when it closes it prevents blood from flowing backward, from the pulmonary artery into the right ventricle.

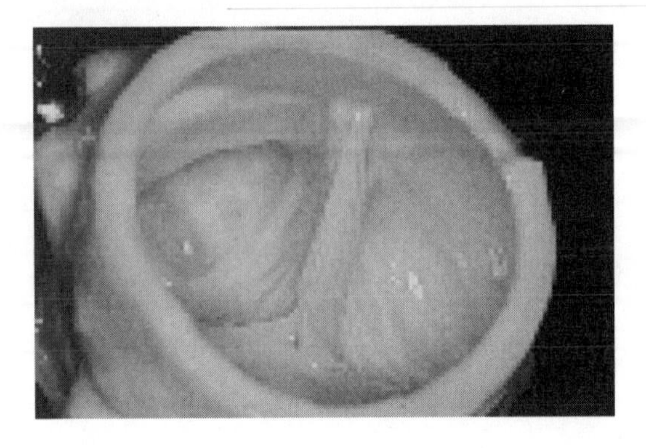

Figure 6.1 Aortic valve. (Reprinted with permission from Lingappa VR, Farey K, Physiological Medicine. New York: McGraw-Hill; 2000.)

6.2 Aortic and Pulmonic Valves

The aortic and pulmonic valves consist of three semilunar cusps that are imbedded within connective tissue. The cusps are attached to a fibrous ring that is embedded in the ventricular septum. The leaflets of the three cusps are lined with endothelial cells and have a dense collagenous core that is adjacent to the aortic side of the leaflets.

A leaflet of an aortic valve is shown in Fig. 6.2. The side of the valve leaflet that is adjacent to the aorta is a fibrous layer within the leaflet, and is called the *fibrosa*. The side of the leaflet that is adjacent to the ventricle is composed of collagen and elastin and is called the *ventricularis*. In a cross section of an aortic leaflet you would see that the ventricularis is thinner than the fibrosa and presents a very smooth and slippery surface to blood flow through the valve. The central portion of the valve, known as the *spongiosa*, contains connective tissue and proteins and is normally not vascularized. The *corpus arantii* (or *nodulus of Arantus*) is a large collagenous mass in the coaptation region, which is thought to aid in valve closure and reduce regurgitation. Coapt literally means to approximate, as at the edge of an closed wound. A coaption

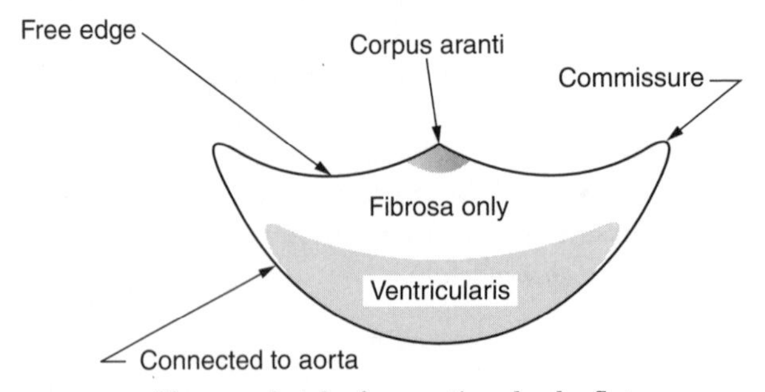

Figure 6.2 Shows a sketch of an aortic valve leaflet.

surface is one that is formed by two overlapping surfaces that approximate one another.

Collagen fibers in the fibrosa and ventricularis are unorganized in the unstressed or unloaded condition. When the valve is loaded, however, the fibers become oriented in a circumferential direction. A fibrous annular ring separates the aorta from the left ventricle.

Three bulges exist at the root of the aorta, immediately superior to the annular ring of the aortic valve. Those three bulges are known as the sinuses of valsalva or the aortic sinuses. Two of the sinuses give rise to the coronary arteries that branch off of the aorta. The right coronary artery comes from the right anterior sinus. The left coronary artery arises from the left posterior sinus. The third sinus is known as the noncoronary sinus or the right posterior sinus.

When the aortic or the pulmonic valve is closed, there is a small overlap of tissue from each leaflet that protrudes from the valve and forms a surface within the aorta. This overlapping surface is once again known as a coaption surface. This overlapping tissue or coaption surface in the aortic or pulmonic valve is known as the lunula. The lunula is important in ensuring that the valve seals to prevent leakage.

The anatomy of the pulmonic valve is similar to that of the aortic valve. The chief difference is that the sinuses are smaller and the annulus is larger in the pulmonic valve. The anatomical dimensions of pulmonic and aortic valves are reported by Yoganathan et al. (1995), in the Biomedical Engineering Handbook. The authors reference a study by Westaby et al. (1984) in which 160 pathologic specimens from cadavers were examined. The mean aortic valve diameter was reported as 23.2 +/− 3.3 mm and the mean pulmonic valve diameter was reported as 24.3 +/− 3.0 mm. Another study that was referenced in the same resource was carried out by Gramiak and Shah in 1970 and used M-mode echocardiography to measure the aortic root diameter. The aortic root diameter at end systole was reported as 35 +/− 4.2 mm and the aortic root diameter at end diastole was reported as 33.7 +/− 4.4 mm.

Diastolic stress aortic valve leaflets has been estimated at 0.25 MPa for a strain of 15 percent by Thubrikar et al. (1993). The strain in the circumferential direction is about 10 percent of the normal systolic length. There is also an increased valve surface area associated with diastole. Recall that the aortic valve is relatively unloaded during systole while the valve is open and blood is flowing through it. Larger stresses and strains occur during diastole when the valve is closed and reverse blood flow is prevented.

The aortic valve is a highly dynamic structure. The leaflets open in 20–30 ms. Blood then accelerates through the valve and reaches its peak velocity after the leaflets are fully open. Peak blood velocity is reached in the first one-third of systole and then the blood begins to decelerate after that peak velocity is reached. During the cardiac cycle, the heart also translates and rotates and the base of the aortic valve varies in size, and moves, mainly along the axis of the aorta. The base perimeter of the aortic valve is largest at the end of diastole, at the time during the cardiac cycle when the valve has been closed the maximum length of time. For an aortic pressure range between 120 and 80 mmHg, the perimeter varies approximately 22 percent.

During systole vortices develop in all three sinuses. These vortices were first described by Leonardo da Vinci (1452–1519) in 1513 and it has been hypothesized that vortices help speed up the closure of the aortic valve. Some of da Vinci's sketches of heart valves are shown in Fig. 6.3. Da Vinci was fascinated by the heart. In his elegant investigation of the valves, he constructed a glass model of the aortic valve and sinuses of valsalva by taking a cast from an ox's heart. Observing the vortices in the sinuses, he concluded that the mechanism of closure of the valves was related to the vortices. When the flow ceased the vortices pushed against one another to help seal the valve. In fact it is true that the vortices create a transverse pressure gradient that pushes leaflets toward the center of the aorta.

In healthy individuals, blood flows through the aortic valve at the beginning of systole and rapidly accelerates to a peak value of approximately 1.35 +/− 0.3 m/s. This value is slightly higher in children, or about 1.5 +/− 0.3 m/s. Pulmonic valve peak velocities are smaller or about 0.75 +/− 0.15 m/s in adults and 0.9 +/− 0.2 m/s in children (Kilner et al. 1993; Weyman 1994).

The aortic valve is prone to acquired as well as congenital heart disease. If blood leaks backwards through the aortic valve, this condition is known as aortic regurgitation. A stenosis is a blockage of blood flow

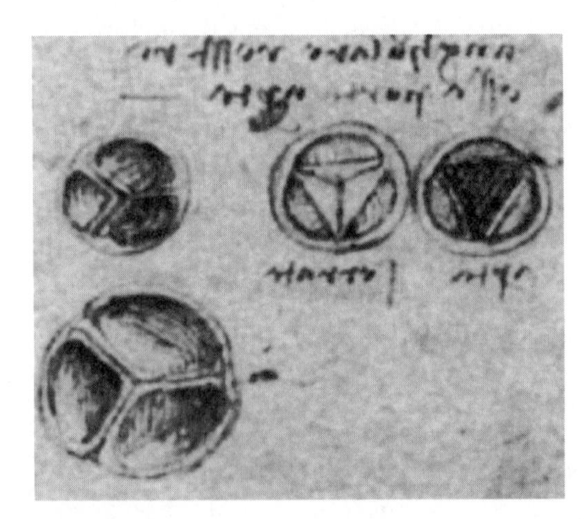

Figure 6.3 Leonardo Da Vinci's sketch of an aortic valve. (Figure from Leonardo Da Vinci, The Anatomy of Man, The Royal Collection © 2005 Her Majesty Queen Elizabeth II).

through the valve. Both aortic regurgitation and stenoses are often caused by the calcification of valve tissue.

6.3 Mitral and Tricuspid Valves

Andreas Vesalius (1514–64) was the Belgian anatomist who was said to have suggested the picturesque term "mitral" to describe the left atrioventricular valve owing to its resemblance to a plan view of the bishop's mitre. A medical view cross-section of the mitral valve is shown in Fig. 6.4. The mitral and tricuspid valves are composed of four primary elements. These elements include the valve annulus, the valve leaflets, the papillary muscles, and the chordae tendineae. The bases of the leaflets form the annulus, which is an elliptical ring of dense collagenous tissue surrounded by muscle. The circumference of the mitral annulus is 8 to 12 cm during ventricular diastole when blood is flowing through the mitral valve from the left atrium to the left ventricle. During ventricular systole the mitral valve closes and the circumference of the mitral annulus is smaller.

Figure 6.5 shows a sketch of a superior view of a mitral valve (the mitral valve as seen from above). The mitral valve is a bicuspid valve and has an aortic (or anterior) leaflet and a mural (or posterior) leaflet. Because of the oblique position of the valve, strictly speaking neither leaflet is anterior or posterior. The aortic leaflet occupies about a third of the annular circumference and the mural leaflet is long and narrow and lines the remainder of the circumference. When the valve is closed the view of the valve from the atrium resembles a smile. Each end of the closure line is referred to as a commissure.

Each leaflet is composed of collagen reinforced endothelium. Striated muscle, nerves fiber, and blood vessels are also present in the mitral

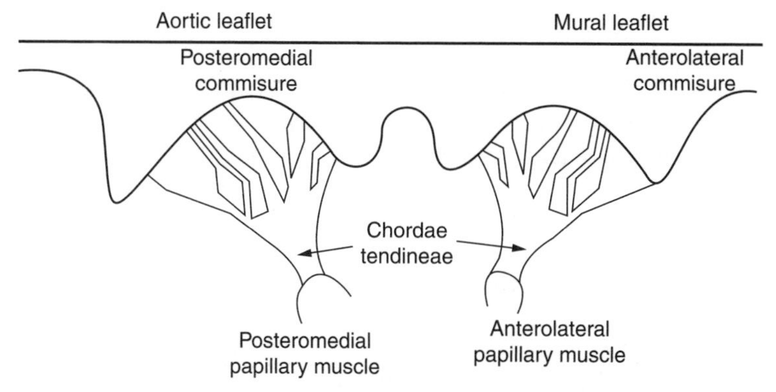

Figure 6.4 Medial view of the dissected mitral valve, cut through the anterolateral commissure.

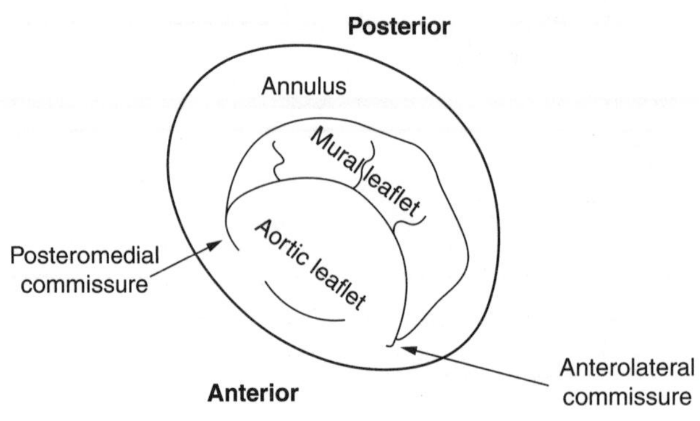

Figure 6.5 Superior view of a mitral valve.

valve. Muscle fibers are usually present on the aortic leaflet but rarely on the mural leaflet.

The aortic (anterior) leaflet is slightly larger than the mural (posterior) leaflet. The combined surface of both leaflets is approximately twice the area of the mitral orifice, allowing for a large area of coaptation permitting effective valve sealing.

The mitral valve is shown in a medial view in Fig. 6.4. The valve has been dissected through the anterolateral commissure and laid open. In this view it can be seen that the anterior and posterior leaflets are not separate entities, but rather one continuous piece of tissue.

The chordae tendineae for both leaflets attach to the papillary muscle. Chordae tendineae consist of an inner core of collagen surrounded by loosely meshed elastin and collagen fibers with an outer layer of endothelial cells. There is an average of about 12 primary chordae tendineae attached to each papillary muscle. The anterolateral and the posteromedial papillary muscle attach to the ventricular wall and tether the mitral valve into place. The chordae tendineae prevent the mitral valve from inverting through the annular ring and into the atrium. That is, in normal physiologic situations they prevent mitral valve prolapse. Improper tethering results in prolapse and mitral regurgitation.

Stresses in mitral valve leaflets of up to 0.22 MPa have been estimated when the left ventricular pressure reaches up to 150 mmHg during ventricular systole. Mitral velocity flow curves show a peak in the curve during early filling. This is known as an E wave. Peak velocities are typically 50 to 80 cm/s at the mitral annulus. A second velocity peak occurs during the atrial systole and is known as an A wave.

In healthy individuals, velocities in the A wave are lower than the velocities during the E wave. Diastolic filling of the ventricle shows a peak inflow during diastolic filling followed by a second peak during atrial systole.

6.4 Clinical Features

Chordae tendineae rupture and papillary muscle paralysis can be consequences of a heart attack. This can lead to bulging of the valve, excessive backward leakage into the atria (regurgitation), and even valve prolapse. Valve prolapse is the condition under which the valve inverts backward into the atrium. Because of these valve problems, the ventricle does not fill efficiently. Significant further damage, and even death, can occur within the first 24 h after a heart attack because of this problem.

6.5 Prosthetic Mechanical Valves

Early mechanical prosthetic heart valves were ball valves. In 1952, Dr. Charles Hufnagel implanted the first ball and cage valve into a patient with aortic valve insufficiency. Although the design of mechanical prosthetic valves has progressed significantly over the past 50 years, ball and cage valves are still the valve of choice by some surgeons.

Björk-Shiley designed the first tilting disk, or single leaflet, valve around 1969. Those valves used a tilting disk occluder that was held in place by a retaining strut. However, the development of the bileaflet valve by St. Jude in 1978, has led to the predominance of this type of bileaflet valve in the mechanical valve market.

One example of a recently approved mechanical, bileaflet aortic valve is the On-X valve (see Fig. 6.6) manufactured by Medical Carbon Research Institute in Austin, Texas. The On-X was approved by the FDA for implant in humans in May 2001. The mitral valve version of the On-X was approved for use by the FDA in March 2002.

The On-X consists of a pyrolytic carbon cage, a Dacron sewing ring fixed by titanium rings, and a bileaflet pyrolytic carbon disk.

Figure 6.6 On-X prosthetic made by MCRI, Austin, Texas (a wholly owned subsidiary of Medical Carbon Research Institute Deutschland Gmbh, Hanover, Germany), with permission.

Pyrolytic carbon is a material used in artificial heart valves because of its mechanical durability and its characteristic ability of not generating blood clots. The material can be said to have low thrombogenicity. Dr Jack Bokros and Dr Vincent Gott discovered pyrolytic carbon in 1966. At that time, pyrolytic carbon was used to coat nuclear fuel particles for gas-cooled nuclear reactors. General Atomics initiated a development project headed by Dr. Bokros to develop a biomedical grade of pyrolytic carbon. The result was Pyrolite, which was incorporated into heart valve designs and remains the most widely used material for heart valves today.

6.5.1 Case study–the Björk-Shiley convexo-concave heart valve

The design and development of the Björk-Shiley convexo-concave heart valve is interesting from a biomedical engineering design as well as from an ethical standpoint. In the early 1960s the earliest human prosthetic heart valves were developed. In 1976, Dr. Viking Björk and Shiley Inc. developed the 60 degree convexo-concave BSCC-60 heart valve. The 60 degree designation indicates that the valve was designed to have a maximum 60 degree opening. The valve had improved flow characteristics over other valves that were on the market at the time. In 1979, Shiley came out with a valve that opened even further. It was the BSCC 70 heart valve. The valve had a cobalt chrome alloy housing with a pyrolytic carbon disk, a Teflon sewing ring and a cobalt chrome retaining strut, which retained the disk in the valve.

During clinical trials of the valve, the first valve failure occurred when the outlet strut fractured. Shiley Inc. reported that failure as an anomaly and the Food and Drug Administration approved the valve for use in the United States. After the BSCC-60 went on the market, outlet strut fractures began to occur. Most of the reported fractures were fatal and between 1980 and 1983 Shiley implemented three voluntary recalls of the valve. During that time, Shiley's share of the prosthetic valve market was up to 60 percent.

In the early 1980s the Food and Drug Administration (FDA) began an investigation of the Björk-Shiley valve. Evidence later surfaced that Shiley Inc. attempted to delay and conceal damaging information about its product. For example, the FDA requested sets of randomly selected, recalled valves for testing. Instead of sending randomly selected valves, Shiley hand picked the valves that were sent to the FDA for testing.

- Case report 1: 53-year-old German female in 1992 in Göttingen.
- Case report 2: 49-year-old Japanese businessman in San Francisco in 1990.

For a case report of a 53-year-old woman who suffered BSVV valve failure, see section 2.2, Clinical Features.

In the end, 89,000 people worldwide received Björk-Shiley convexo-concave (BSCC) valves. By November 1991, 466 outlet strut fractures had been reported. A large follow-up study of 60 or 70 degree convexo-concave valves, with 2303 valves implanted in the Netherlands, was published in 1992 (van der Graaf). It was determined that most patients who died with a BSCC valve had died outside the hospital. Only 24 percent of the patients who had died with the BSCC valves implanted had been examined for strut failure. In the study it was determined that the cumulative failure rate for BSCC60 valves was 4.2 percent over 8 years. The cumulative failure rate for BSCC70 valves was 17.4 percent over 8 years. When a strut fracture occurs in a BSCC heart valve, the mortality rate is very high. For aortic valve with strut failure, 85 percent of the patients die. The overall mortality rate for strut fractures is 75 percent. Shiley Inc. voluntarily withdrew the BSCC heart valve from the market in 1986.

In a study published in Lancet (de Mol, 1994), 24 valves were replaced in 22 patients who were considered to be at high risk with respect to failure of a BSCC heart valve. Seven of the 24 valves had experienced a single leg fracture of the strut. The authors wrote, "This finding supports our hypothesis that, owing to the lethal character of the failing aortic-valve strut, fractures remain under-reported, few such patients reach hospital and necropsy is rarely done."

In a report by Steyerberg (January 2000), it was estimated that 35,000 patients were still alive with one or more implanted BSCC valves. Pfizer purchased Shiley in 1979 at the onset of its Convexo-Concave valve ordeal. In 1992, after years of litigation, Pfizer sold Shiley's businesses to Italy's Sorin Biomedical. Sorin opted not to purchase rights to the C/C valve. Later, a firm by the name of Alliance Medical Technologies, comprised of former Shiley and Pfizer employees, purchased the rights to Shiley's Monostrut heart valve line from Sorin. The Monostrut is still available outside the United States. It has developed a good reputation for reliability and good hemodynamics. There are no welded struts on the Monostrut valve as both the inlet and outlet strut are both part of the same piece of metal and the valve ring.

Patients with an implanted Björk-Shiley BSCC heart valve seem to be limited to just a couple of options.

1. They can elect to keep the valve and hope it does not fail. So far the published failure rates are in the 2 to 17 percent range. The risk of failure is thought to be dependent on fatigue and therefore cumulative with time. In the long run, the failure rates may be much higher.
2. Patients may elect preventive valve replacement surgery. The risk of death for a 50-year-old male from valve replacement surgery is about

5 percent. The cost of such a replacement surgery is conservatively in excess of $ 25,000 in the United States.

In January 1988, the following letter from the FDA was sent to physicians to provide information about the risk of outlet strut fractures.

> January 1998
> Dear Doctor:
> This letter provides new information about the risk of outlet strut fracture for 60 degree Björk-Shiley Convexo-Concave (BSCC) heart valves and new recommendations from an independent expert panel regarding prophylactic valve replacement. The recommendations are described in detail in the enclosed attachments.
> Under the Settlement Agreement that was entered into by a worldwide class of BSCC valve patients and Shiley Incorporated and approved by the U.S. District Court in Cincinnati, Ohio in Bowling v. Pfizer, an independent expert medical and scientific panel consisting of cardiothoracic surgeons, cardiologists, and epidemiologists was created called the Supervisory Panel. Under the terms of the Settlement Agreement, the Supervisory Panel is charged with the responsibilities of conducting studies and research, and of making recommendations regarding which BSCC heart valve patients should be considered for prophylactic valve replacement. The Supervisory Panel's recommendations also serve to determine which class members qualify for explantation benefits under the Settlement Agreement. The Panel's work has enabled it to develop these present guidelines for valve replacement surgery. The Panel's guidelines represent a departure from past reliance upon the valve replacement surgery guidelines developed by a Medical Advisory Panel created by Shiley Incorporated.
>
> Letter from the FDA website:
> http://www.fda.gov/medwatch/safety/1998/bjork.htm

By 2004 van Gorp published another study in which he investigated all 2263 Dutch BS Convexo-Concave patients. For the surviving patients in 1992 (n = 1330), they calculated the expected differences in life expectancy with and without valve replacement. Of 1330 patients, 96 (7 percent) had undergone valve replacement. The investigators concluded that 117 patients (9 percent) had an estimated gain in life expectancy after valve replacement.

6.6 Prosthetic Tissue Valves

One disadvantage to a mechanical prosthetic valve is that the patient is required to undergo lifelong anticoagulation therapy. Although pyrolytic carbon has low thrombogenicity, valves made from this material still generate enough blood clots to warrant this type of therapy.

Naturally occurring heart valves avoid the blood clotting problem as well as having better hemodynamic properties.

Homografts are valves that come from a different member of the same species. These allografts typical come from cadavers and therefore consist of nonliving tissue. The advantage to containing nonliving cells is that the valves will not be rejected by the immune system of the host. The disadvantage is that they do not have cellular regeneration associated with living tissue and therefore are susceptible to long term mechanical wear and damage. Human allografts are also available in only very limited supplies. One interesting alternative approach to cadaveric allografts that is sometimes used is to transplant a patients own pulmonary valve into the aortic position.

One alternative approach to an allograft is a xenograft. A xenograft is an implant derived from another species. In 1969, Kaiser and coworkers described a tissue valve substitute that consisted of a glutaraldehyde treated, explanted porcine valve. This porcine valve became commercially available in 1970 as the Hancock porcine xenograft.

Tissue valves are rarely used in young children and young adults because of valve leaflet calcification and a relatively shorter valve life when compared to mechanical valves.

Bibliography

de Mol BA, Kallewaard M, McLellan RB, van Herwerden LA, Defauw JJ, van der Graaf Y. Single-leg strut fractures in explanted Bjork-Shiley valves, *Lancet*. 1994 January; 343(8888):9–12.

Hajar R. Art and Medicine, *Da Vinci's Anatomical Drawings The Heart*. 2002 June–August; Vol 3(2):131–133.

Ho S. Anatomy of the Mitral Valve, *Heart*. 2002 November; 88(4):5–10.

Kilner PJ, Yang GZ, Mohiaddin RH, et al. Helical and retrograde secondary flow patterns in the aortic arch studied by three-directional magnetic resonance velocity mapping, *Circulation*. 1993;88(part I):2235.

Steyerberg EW, Kallewaard M, van der Graaf Y, van Herwerden LA, Habbema JD. Decision analyses for prophylactic replacement of the Bjork-Shiley convexo-concave heart valve: an evaluation of assumptions and estimates, *Med Decis Making*. 2000 Jan-Mar; 20(1):20–32.

Thubrikar M, Heckman JL, Nolan SP. High speed cine-radiographic study of aortic valve motion, *J Heart Valve Dis*. 1993;2:653.

van der Graaf Y, de Waard F, van Herwerden LA, Defauw J. Risk of strut fracture of Bjork-Shiley valves, *Lancet*. 1992 Feb 1;339(8788):257–61.

van Gorp MJ, Steyerberg EW, Van der Graaf Y. Decision guidelines for prophylactic replacement of Bjork-Shiley convexo-concave heart valves: impact on clinical practice, *Circulation*. 2004 May 4;109(17):2092–6.

Westaby S, Karp R B, Blackstone EH, et al. Adult human valve dimensions and their surgical significance, *Am J Cardiol*. 1984;53:552.

Weyman AE. *Principles and Practices of Echocardiography*. Philadelphia, PA. Lea & Febiger; 1994.

Yoganathan AP, Hopmeyer J, Heinrich R S, *Biomedical Engineering Handbook*, Joseph D. Bronzino, ed. Boca Raton, FL. CRC Press; 1995.

Chapter

7

Pulsatile Flow in Large Arteries

7.1 Fluid Kinematics

Let us begin with the mathematical description of the motion of the fluid elements moving in a flow field. It will be convenient to express the velocity in terms of three Cartesian components so that velocity becomes a function of x, y, and z spatial coordinates, as well as a function of time. Also u, v, and w are the velocity components in the x, y, and z directions, respectively. Written in a concise form:

$$\text{Velocity} = \vec{V}(x,y,z,t)$$
$$\vec{V} = u\hat{i} + v\hat{j} + w\hat{k} = u(x,y,z,t)\hat{i} + v(x,y,z,t)\hat{j} + w(x,y,z,t)\hat{k} \tag{7.1}$$

The corresponding expressions for acceleration are shown in Eqs. (7.2) through (7.5). Note that Eq. (7.2) is the vector representation of the acceleration and Eqs. (7.3) through (7.5) show the three scalar Cartesian components of the acceleration.

$$\text{Acceleration} = \vec{a}(x,y,z,t) = \frac{\partial \vec{V}}{\partial t} + u\frac{\partial \vec{V}}{\partial x} + v\frac{\partial \vec{V}}{\partial y} + w\frac{\partial \vec{V}}{\partial z} \tag{7.2}$$

$$a_x = \frac{\partial u}{\partial t} + u\frac{\partial u}{\partial x} + v\frac{\partial u}{\partial y} + w\frac{\partial u}{\partial z} \tag{7.3}$$

$$a_y = \frac{\partial v}{\partial t} + u\frac{\partial v}{\partial x} + v\frac{\partial v}{\partial y} + w\frac{\partial v}{\partial z} \tag{7.4}$$

$$a_w = \frac{\partial w}{\partial t} + u\frac{\partial w}{\partial x} + v\frac{\partial w}{\partial y} + w\frac{\partial w}{\partial z} \tag{7.5}$$

It is also possible to more concisely express the acceleration using the operator $D(\,)/Dt$, which is known as the material derivative or substantial derivative. The parenthesis in this notation encloses a vector or scalar to be named later, for example, the velocity vector $\vec{V}$ in Eq. (7.6).

$$\vec{a} = \frac{D\vec{V}}{Dt} \tag{7.6}$$

The operator $D(\,)/Dt$ is defined in Eq. (7.7).

$$\frac{D(\,)}{Dt} = \frac{\partial(\,)}{\partial t} + u\frac{\partial(\,)}{\partial x} + v\frac{\partial(\,)}{\partial y} + w\frac{\partial(\,)}{\partial z} \tag{7.7}$$

It is perhaps valuable here to point out that the total material derivative $D(\vec{V})/Dt$ has two parts. One part is the local time derivative and the other comes from the fact that our fluid particle convects into a region of different velocity. Therefore a particle in steady flow can accelerate. This conception is fundamental to an Eulerian formulation.

It is also useful here to define the gradient operator $\nabla(\,)$ and rewrite the material derivative more concisely.

$$\vec{\nabla}(\,) = \frac{\partial(\,)}{\partial x}\hat{i} + \frac{\partial(\,)}{\partial y}\hat{j} + \frac{\partial(\,)}{\partial z}\hat{k} \tag{7.8}$$

$$\frac{D(\,)}{Dt} = \frac{\partial(\,)}{\partial t} + \vec{V}\cdot\vec{\nabla} \tag{7.9}$$

Equation (7.10) also introduces the Laplacian operator. The Laplacian yields a scalar, by taking the dot product of the vector gradient operator $\vec{\nabla}$ with itself as shown in Eq. (7.11).
Laplacian operator (a scalar)

$$\nabla^2(\,) = \vec{\nabla}(\,)\cdot\vec{\nabla}(\,) \tag{7.10}$$

$$\nabla^2(\,) = \frac{\partial^2(\,)}{\partial x^2} + \frac{\partial^2(\,)}{\partial y^2} + \frac{\partial^2(\,)}{\partial z^2} \tag{7.11}$$

7.2 Continuity

The continuity equation, or conservation of mass, was introduced in Chap. 1 of this book. It was written as $\rho_1 A_1 V_1 = \rho_2 A_2 V_2 =$ constant, where ρ is fluid density, A is the cross-sectional area of the blood vessel and V is the average blood velocity across the cross section.

Since we are now dealing with the flow at a point the more general version of the conservation of mass equation, which is also commonly referred to as the continuity equation can be written by

$$\frac{\partial\rho}{\partial t} + \frac{\partial(\rho u)}{\partial x} + \frac{\partial(\rho v)}{\partial y} + \frac{\partial(\rho w)}{\partial z} = 0 \tag{7.12}$$

The continuity equation is one of the basic governing equations of fluid mechanics and is valid for compressible and incompressible flows in the form written before. It is also valid for steady as well as pulsatile flows in that form.

For the case of incompressible flows, like blood, the continuity equation can also be simplified to

$$\frac{\partial(u)}{\partial x} + \frac{\partial(v)}{\partial y} + \frac{\partial(w)}{\partial z} = 0 \tag{7.13}$$

7.3 Complex Numbers

If one uses polynomials to model blood flow, then complex numbers are useful. Complex numbers use a number called j (or i in some disciplines), with the definition that $j^2 = -1$. These numbers are written in the form of a vector where $\vec{z} = a + jb$, where a and b are real numbers. The part a is designated the real part of the complex number, $a = \mathrm{Re}(\vec{z})$, and b is the imaginary part, $b = \mathrm{Im}(\vec{z})$. Leonhard Euler (1707–1783) made the observation that

$$e^{jx} = \cos x + j \sin x \tag{7.14}$$

Euler's observation became very useful in the solution to differential equations since it allows one to interpret the exponential of a complex imaginary number in terms of sines and cosines.

The modern convention is to place the real part of the number on the horizontal x-axis and to display the imaginary part on the y-axis. In this way numbers like $\vec{z} = a + j\,b$ are displayed in the complex plane. See Fig. 7.1 which shows a vector $\vec{z}$ represented in the complex plane.

For this book I will use Euler's identity in the following form:

$$e^{j\omega t} = \cos(\omega t) + j \sin(\omega t) \tag{7.15}$$

where $j^2 = -1$
$j^3 = -j$
t = time
ω = fundamental frequency of the signal
$\cos(\omega t) = \mathrm{Re}[e^{j\omega t}]$ or the real part
$\sin(\omega t) = \mathrm{Im}[e^{j\omega t}]$ or the imaginary part

In the type of fluid mechanics problem with which we are dealing in this book, t represents the independent variable time. There is also a fundamental frequency associated with the problem, which is typically the heart rate in rad/s. It is also possible to represent ωt as θ, now with units of radians.

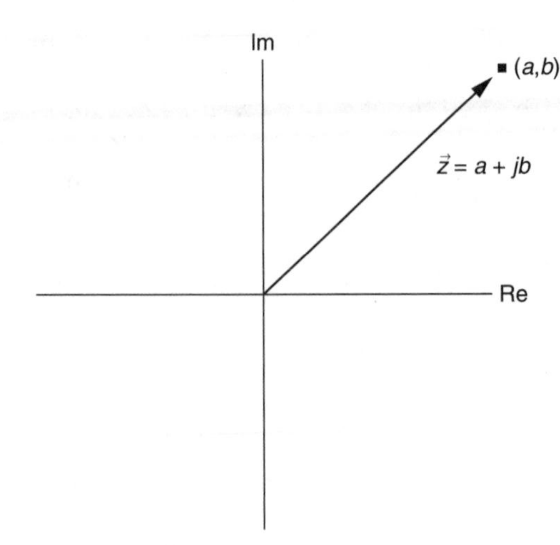

Figure 7.1 The vector $\vec{z} = a + jb$ in the complex plane.

Now that we have this useful concept of complex numbers, it is possible to add, subtract, multiply, and divide the numbers. For example, to add complex numbers it is necessary to add the real parts and the imaginary parts separately.

$$(A + Bj) + (C + Dj) = (A + C) + (B + D)j$$

For multiplication it is necessary to convert to the exponential form with a magnitude and an angle and to multiply the magnitudes and add the angles.

$$\vec{A}\vec{B} = (Ae^{j\theta})(Be^{j\phi}) = ABe^{j(\theta+\phi)}$$

The complex conjugate of $\vec{z} = a + j\,b$ is defined as $\vec{z}^{\,*} = a - jb$. The square root of a complex number $\vec{z}$ times its complex conjugate results in the magnitude of that number. The magnitude of $\vec{z}$ is $\sqrt{a^2 + b^2}$ which could also be written as $\sqrt{(\vec{z})\cdot(\vec{z}^{\,*})}$. The argument of $\vec{z}$ is $\tan^{-1}(b/a)$.

Example Problem

$$\vec{A} = (3 + 4j)$$

The magnitude of A is written as $|\vec{A}| = \sqrt{3^2 + 4^2} = 5$. The angle θ is written as $\theta = \tan^{-1}(4/3) = 53.1^\circ$.

So an alternate form of writing the complex number in phasor notation is $\vec{A} = 5\angle 53.1^{\circ}$.

Example Problem

A second example of a complex number is $\vec{B} = 5 + 12j$. Now the magnitude of B is $|\vec{B}| = \sqrt{5^2 + 12^2} = 13$ and the angle ϕ is $\phi = \tan^{-1}(12/5) = 67.4^{\circ}$. Finally the phasor notation is $\vec{B} = 13\angle 67.4^{\circ}$.

Now from our two complex number examples, it is possible to show a multiplication example:

$$\vec{A}\vec{B} = (5e^{j53.1})(13e^{j67.4}) = 65e^{j(120.5)}$$

$$65e^{j(120.5)} = 65\,[\cos(120.5) + j\sin(120.5)] = -33 + 56j$$

From the complex number $\vec{A}$ and $\vec{B}$ the division example is written:

$$\vec{A}/\vec{B} = (5e^{j53.1})/(13e^{j67.4}) = \frac{5}{13}e^{j(53.1-67.4)} = \frac{5}{13}e^{j(-14.3)}$$

$$= \frac{5}{13}[\cos(-14.3) + j\sin(-14.3)] = 0.373 + -0.0947j$$

7.4 Fourier Series Representation

The Fourier series theorem that was published in 1822 states:

> *A periodic function f(t) with a period T can be represented by the sum of a constant term, a fundamental of period T, and its harmonics.*

In general a Fourier series can be used to expand any periodic function into an infinite sum of sines and cosines. This is an extremely useful way to break up an arbitrary periodic function into a set of simple terms.

For the case of pulsatile flow, where u is velocity and P is pressure, the partial derivative of velocity u with respect to time is not zero. In addition, the partial derivative of pressure P with respect to distance along the tube, x, is also nonzero. These were important conditions for Poiseuille's law. Therefore Poiseuille flow is no longer a good estimate for the case of pulsatile flow. Written more concisely:

$$\frac{\partial u}{\partial t} \neq 0 \qquad \text{and} \qquad \frac{\partial P}{\partial x} \neq 0$$

Since the pressure waveform is periodic, it is convenient to write the partial derivative of pressure $\partial P/\partial x$ using a Fourier series. Figure 7.2 shows

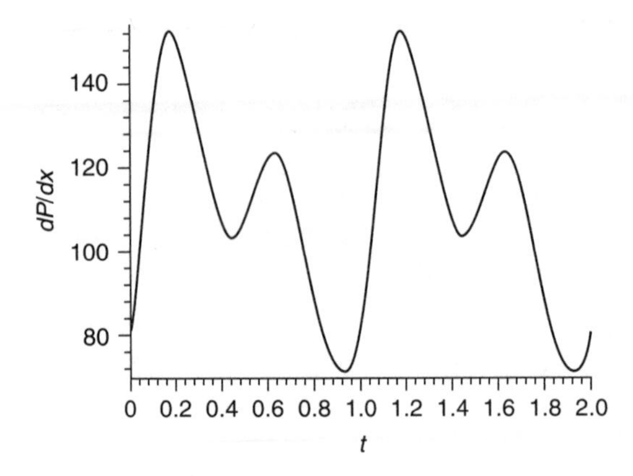

Figure 7.2 Plot of a pulsatile pressure waveform showing pressure drop versus time for a pulsatile flow condition.

a plot of a pulsatile pressure waveform that shows pressure drop versus time for a pulsatile flow condition.

This periodic function depends on the fundamental frequency of the signal, ω (heart rate in rad/s), and the time t. We can write any such function as a sum of sine and cosine terms, with appropriate coefficients, known as Fourier coefficients.

Since $\partial P/\partial x$ is periodic, it may be written as

$$\frac{\partial P}{\partial x} = A_0 + A_1\cos(\omega t) + A_2\cos(2\omega t) + A_3\cos(3\omega t) + \cdots + B_1\sin(\omega t) + B_2\sin(2\omega t) + B_3\sin(3\omega t) + \cdots \quad (7.16)$$

This is a Fourier series representation of the pressure gradient. If we have a periodic pressure waveform, it is possible to obtain the Fourier coefficients, $A_1, A_2, A_3, \ldots B_1, B_2, B_3, \ldots$ by a number of common methods. If a set of digitized data exists for example, a fast Fourier transform implemented in many types of software will return the set of Fourier coefficients.

It is also possible to write the Fourier series using complex numbers by using Euler's identity to convert the sines and cosines to exponentials. The equivalent expression would be

$$\frac{\partial P}{\partial x} = \mathrm{Re}\left[\sum_{n=0}^{\infty} a_n e^{j\omega nt}\right] \quad (7.17)$$

where $a_n = A_n - B_n j$, n is the number of the harmonic
ω = fundamental frequency (rad/s)
A_0 = mean pressure gradient (nonpulsatile)

Note: $a_n = (A_n - B_n j)$ or $a_n = \text{conjugate}\,(A_n + B_n j)$

7.5 Navier-Stokes Equations

The Navier-Stokes equations are named in honor of the French mathematician L. M. H. Navier (1758–1836) and the English mechanician Sir G. G. Stokes (1819–1903), who were responsible for their formulation. These equations are nonlinear, second order, partial differential equations and they are the considered to be the governing differential equations of motion for incompressible, newtonian fluids.

The Navier-Stokes equations can now be written rather efficiently in the following form:

$$\rho \frac{D\vec{V}}{Dt} = \rho \vec{g} - \vec{\nabla} P + \mu \nabla^2 \vec{V} \tag{7.18}$$

In Eq. (7.18), r is the fluid density, $D\vec{V}/D$ is the material derivative of the velocity vector, $\vec{g}$ is gravitational acceleration, $\vec{\nabla}P$ represents the pressure gradient, and μ represents viscosity.

The left-hand side of Eq. (7.18), $\rho\ D\vec{V}/Dt$, represents something analogous to a mass (density, ρ) times an acceleration (substantial derivative of the velocity vector) and can be expanded to the following three scalar equations:

$$a_x = \underbrace{\rho\frac{\partial(u)}{\partial t}}_{\substack{\text{transient}\\ \text{inertia}\\ \text{term}}} + \underbrace{\rho\left(\overbrace{u\frac{\partial(u)}{\partial x}}^{\substack{\text{nonlinear}\\ \text{in dependent variable}}} + v\frac{\partial(u)}{\partial y} + w\frac{\partial(u)}{\partial z} \right)}_{\substack{\text{convective}\\ \text{inertia}\\ \text{term}}} \tag{7.19}$$

$$a_y = \underbrace{\rho\frac{\partial(v)}{\partial t}}_{\substack{\text{transient}\\ \text{inertia}\\ \text{term}}} + \underbrace{\rho\left(\overbrace{u\frac{\partial(v)}{\partial x}}^{\substack{\text{nonlinear}\\ \text{in dependent variable}}} + v\frac{\partial(v)}{\partial y} + w\frac{\partial(v)}{\partial z} \right)}_{\substack{\text{convective}\\ \text{inertia}\\ \text{term}}} \tag{7.20}$$

$$a_z = \underbrace{\rho\frac{\partial(w)}{\partial t}}_{\substack{\text{transient}\\ \text{inertia}\\ \text{term}}} + \underbrace{\rho\left(\overbrace{u\frac{\partial(w)}{\partial x}}^{\substack{\text{nonlinear}\\ \text{in dependent variable}}} + v\frac{\partial(w)}{\partial y} + w\frac{\partial(w)}{\partial z} \right)}_{\substack{\text{convective}\\ \text{inertia}\\ \text{term}}} \tag{7.21}$$

Note that Eqs. (7.19) through (7.20) are formulated on a per volume basis. The reader can observe, for example, that $\partial P/\partial x$ has units of $F/L^2/L = F/L^3$ = force/volume. One might wonder what kind of forces are causing all of that mass to accelerate. The right-hand side of the Navier-Stokes equations provides insight into that question. The right-hand side can be expanded into the three scalar expressions shown as follows:

$$\underbrace{\rho g_x}_{\text{weight}} \underbrace{- \frac{\partial P}{\partial x}}_{\text{pressure force}} + \underbrace{\mu\left(\frac{\partial^2(u)}{\partial x^2} + \frac{\partial^2(u)}{\partial y^2} + \overbrace{\frac{\partial^2(u)}{\partial z^2}}^{\text{higher ordered derivative, 2nd order PDE}}\right)}_{\text{viscous force}} \tag{7.22}$$

$$\underbrace{\rho g_y}_{\text{weight}} \underbrace{- \frac{\partial P}{\partial y}}_{\text{pressure force}} + \underbrace{\mu\left(\frac{\partial^2(v)}{\partial x^2} + \frac{\partial^2(v)}{\partial y^2} + \overbrace{\frac{\partial^2(v)}{\partial z^2}}^{\text{higher ordered derivative, 2nd order PDE}}\right)}_{\text{viscous force}} \tag{7.23}$$

$$\underbrace{\rho g_z}_{\text{weight}} \underbrace{- \frac{\partial P}{\partial z}}_{\text{pressure force}} + \underbrace{\mu\left(\frac{\partial^2(w)}{\partial x^2} + \frac{\partial^2(w)}{\partial y^2} + \overbrace{\frac{\partial^2(w)}{\partial z^2}}^{\text{higher ordered derivative, 2nd order PDE}}\right)}_{\text{viscous force}} \tag{7.24}$$

The left-hand side of the equation could also be written in terms of cylindrical components. For blood flow problems in which the blood is flowing in cylindrical arteries, this type of coordinate system is convenient. The expression for the x-components of inertia (axial direction) using cylindrical components is shown as follows:

$$\underbrace{\rho\frac{\partial(u)}{\partial t}}_{\text{transient inertia term}} + \underbrace{\rho\left(u\overbrace{\frac{\partial(u)}{\partial x}}^{\text{nonlinear in dependent variable}} + v_r\frac{\partial(u)}{\partial r} + \frac{v_\theta}{r}\frac{\partial(u)}{\partial \theta}\right)}_{\text{convective inertia term}} \tag{7.25}$$

The right-hand side or the Navier-Stokes equation written in cylindrical components for the axial component then completes the equation.

$$\underbrace{\rho\frac{\partial(u)}{\partial t}}_{\text{transient inertia term}} + \underbrace{\rho\left(u\overbrace{\frac{\partial(u)}{\partial x}}^{\text{nonlinear in dependent variable}} + v_r\frac{\partial(u)}{\partial r} + \frac{v_\theta}{r}\frac{\partial(u)}{\partial \theta}\right)}_{\text{convective inertia term}}$$

$$= \rho g_x - \frac{\partial P}{\partial x} + \mu\left(\frac{\partial^2(u)}{\partial x^2} + \frac{\partial^2(u)}{\partial r^2} + \frac{1}{r}\frac{\partial(u)}{\partial r} + \frac{1}{r^2}\frac{\partial^2(u)}{\partial \theta^2}\right) \tag{7.26}$$

Example Problem

Show that the parabolic velocity profile from horizontal Poiseuille flow is a solution of the Navier-Stokes equations using the cylindrical form.

Solution:

For Poiseuille flow we have already assumed:

1. Steady flow
2. Uniform flow (constant cross section)
3. Laminar flow
4. Axially symmetric pipe flow

From Eq. (1.11) in Chap. 1, we saw that the parabolic velocity profile from Poiseuille flow has the form of

$$u = \frac{1}{4\mu}\frac{dP}{dx}[r^2 - R^2]$$

where u is the velocity component along the tube as a function of r, μ is fluid viscosity (a constant for newtonian fluids), $\partial P/\partial x$ is the pressure gradient along the tube (a constant for Poiseuille flow), r is the radius variable, and R is the radius of the tube (a constant).

The Navier-Stokes equation for the component along the axial direction can be written from Eqs. (7.25) and (7.26).

$$\rho\frac{\partial(u)}{\partial t} + \rho\left(u\frac{\partial(u)}{\partial x} + v_r\frac{\partial(u)}{\partial r} + \frac{v_\theta}{r}\frac{\partial(u)}{\partial\theta}\right)$$
$$= \rho g_x - \frac{\partial P}{\partial x} + \mu\left(\frac{\partial^2(u)}{\partial x^2} + \frac{\partial^2(u)}{\partial r^2} + \frac{1}{r}\frac{\partial(u)}{\partial r} + \frac{1}{r^2}\frac{\partial^2(u)}{\partial\theta^2}\right) \quad (7.27)$$

For Poiseuille flow we have already assumed steady, uniform, laminar, axially symmetric, pipe flow. Therefore, there is no change in velocity with respect to time, so $\partial u/\partial t = 0$. Since the flow is axially symmetric there is no swirling flow and no velocity in the radial or transverse direction and therefore no change in velocity in either the radial or transverse direction so $\partial u/\partial\theta = 0$ and $v_r = 0$ and $v_\theta = 0$. Since Poiseuille flow is uniform, there is no change in velocity u in the x (axial) direction, so $\partial u/\partial x = 0$. The equation has now become somewhat more simplified. Equation (7.28) tells us that for Poiseuille flow, there are no accelerations and that viscous forces must balance the pressure forces and gravitational forces.

$$0 = \rho g_x - \frac{\partial P}{\partial x} + \mu\left(\frac{\partial^2(u)}{\partial x^2} + \frac{\partial^2(u)}{\partial r^2} + \frac{1}{r}\frac{\partial(u)}{\partial r} + \frac{1}{r^2}\frac{\partial^2(u)}{\partial\theta^2}\right) \quad (7.28)$$

Since this flow is also horizontal and as already mentioned u does not vary in the x-direction or the θ direction, Eq. (7.28) further simplifies to

$$\frac{\partial P}{\partial x} = \mu\left(\frac{\partial^2(u)}{\partial r^2} + \frac{1}{r}\frac{\partial(u)}{\partial r}\right) \tag{7.29}$$

Since the proposed solution to the differential in Eq. (7.29) is

$$u = \frac{1}{4\mu}\frac{dP}{dx}[r^2 - R^2]$$

then we need to find the first and second derivative of u with respect to r and plug it into the equation to check for equality. We should also recognize explicitly that $dP/dx = \partial P/\partial x$ since the pressure under Poiseuille flow conditions is only a function of x and does not vary with time.

$$\frac{\partial u}{\partial r} = \frac{2}{4\mu}\frac{dP}{dx}r \tag{7.30}$$

$$\frac{\partial^2 u}{\partial r^2} = \frac{2}{4\mu}\frac{dP}{dx} \tag{7.31}$$

$$\frac{\partial P}{\partial x} = \mu\left[\left(\frac{2}{4\mu}\frac{dP}{dx}\right) + \frac{1}{r}\left(\frac{2}{4\mu}\frac{dP}{dx}r\right)\right] \tag{7.32}$$

The equality holds true showing that the Poiseuille flow parabolic velocity profile is one simple solution of the Navier-Stokes equations.

7.6 Pulsatile Flow in Rigid Tubes: Wormersley Solution

Consider the following problem. Given a pressure waveform and the geometry of the artery and some fluid characteristics, estimate the flow rate in terms of volume/time. This may be a pulsatile pressure waveform as is seen in the relatively larger arteries in humans. The blood flow rate is a measure of perfusion, or simply said, how much oxygen may be provided to the downstream, or distal tissues. Therefore, a mathematical model of flow rate under pulsatile conditions can be a very useful device.

To estimate flow from a pulsatile driving pressure in rigid tubes, we will begin by assuming a newtonian fluid, uniform, laminar, axially symmetric, pipe flow. This is similar to the Poiseuille flow problem, but now we are considering pulsatile flow rather than steady.

We will need to have a driving function for the pressure. Recall from Eq. (7.17) that

$$\frac{\partial P}{\partial x} = \mathrm{Re}\left[\sum_{n=0}^{\infty} a_n e^{\,j\omega nt}\right]$$

Therefore for each harmonic n we can write each component of the driving pressure by using Eq. (7.33).

$$\left.\frac{\partial P}{\partial x}\right|_n = \left[a_n e^{\,j\omega nt}\right] \tag{7.33}$$

For each component of the driving pressure from 0 to n, it is also possible to write the Navier-Stokes equation given in Eq. (7.27). The use of the subscript n emphasizes that we are solving for one generic nth, component and then, at the end, assembling the total solution as a sum of components.

$$\rho\frac{\partial(u)}{\partial t} + \rho\left(u\frac{\partial(u)}{\partial x} + v_r\frac{\partial(u)}{\partial r} + \frac{v_\theta}{r}\frac{\partial(u)}{\partial\theta}\right) = \rho g_x - \frac{\partial P}{\partial x} + \mu\left(\frac{\partial^2(u)}{\partial x^2} + \frac{\partial^2(u)}{\partial r^2} + \frac{1}{r}\frac{\partial(u)}{\partial r} + \frac{1}{r^2}\frac{\partial^2(u)}{\partial\theta^2}\right)$$

For this pulsatile flow case we have assumed steady, uniform, laminar, axially symmetric, pipe flow. Since the flow is axially symmetric there is no swirling flow and no velocity in the radial or transverse directions and therefore no change in velocity in either the radial or transverse direction so $\partial u/\partial\theta = 0$ and $v_r = 0$ and $v_\theta = 0$. Since the flow is uniform, there is no change in velocity u in the x (axial) direction so $\partial u/\partial x = 0$. The flow is also horizontal therefore Eq. (7.27) simplifies to

$$\frac{a_n e^{\,j\omega nt}}{\rho} = \nu\frac{\partial^2(u_n)}{\partial r^2} + \frac{\nu}{r}\frac{\partial(u_n)}{\partial r} - \frac{\partial(u_n)}{\partial t} \tag{7.34}$$

This is a linear, second order, partial differential equation (PDE) with a driving function. One possible solution to the PDE is

$$u_n = \underbrace{f_n(r)}_{f\text{ is not time dependent}}\quad e^{\,j\omega nt} \tag{7.35}$$

In order to try this solution in Eq. (7.34), we will need to have the derivative of u with respect to time, t, and the derivate of u with respect to r and also the second derivative of u with respect to r.

$$\frac{du_n}{dr} = \frac{df_n(r)}{dr}e^{j\omega nt} \tag{7.36}$$

$$\frac{d^2u_n}{dr^2} = \frac{d^2f_n(r)}{dr^2}e^{j\omega nt} \tag{7.37}$$

$$\frac{du_n}{dt} = j\omega n f_n(r)e^{j\omega nt} \tag{7.38}$$

Now by using our proposed solution to the differential equation, it is possible to rewrite Eq. (7.34) in the following manner so that an exponential appears in each term.

$$\frac{a_n e^{j\omega nt}}{\rho} = \nu \frac{d^2 f_n(r)}{dr^2} e^{j\omega nt} + \frac{\nu}{r} \frac{df_n(r)}{dr} e^{j\omega nt} - j\omega n f_n(r) e^{j\omega nt} \tag{7.39}$$

Now it is possible to divide Eq. (7.39) by $e^{j\omega nt}$ to remove the time dependence. The result would be

$$\frac{a_n}{\rho} = \nu \frac{d^2 f_n(r)}{dr^2} + \frac{\nu}{r} \frac{df_n(r)}{dr} - j\omega n f_n(r) \tag{7.40}$$

This allows us to treat time dependence and spatial dependence separately. Also we can replace n with m/r and $-j$ with j^3 to obtain Eq. (7.41).

$$\frac{a_n}{\mu} = \frac{d^2 f_n(r)}{dr^2} + \frac{1}{r} \frac{df_n(r)}{dr} + \frac{j^3 \omega n f_n(r)}{\nu} \tag{7.41}$$

Now we have an ordinary differential equation instead of a partial differential equation and one that is not time dependent. Equation (7.41) looks a little bit like a zero-order Bessel differential equation. The homogenous differential equation for the case without a driving function would be

$$\frac{d^2 f_n(r)}{dr^2} + \frac{1}{r} \frac{df_n(r)}{dr} + \frac{j^3 \omega n f_n(r)}{\nu} = 0 \tag{7.42}$$

A zero-order Bessel differential equation has the form

$$\frac{d^2 g}{ds^2} + \frac{1}{s} \frac{dg}{ds} + g = 0 \tag{7.43}$$

The solution to Eq. (7.42) is a Bessel function of order zero and complex arguments, which is well known and arises in problems connected with the distribution of current in conductors of finite size. One homogenous solution is

$$f_n(r) = c_1 J_0(\lambda r)[+ c_2 K_0(\lambda r) \text{ in general}] \tag{7.44}$$

where c_1 = constant
J_0 = zero-order Bessel function
λ is defined in Eq. (7.45)
J_0 is given by Eq. (7.46)

$$\lambda^2 = \frac{j^3 \omega n}{\nu} \tag{7.45}$$

$$J_0(x) = 1 - \frac{x^2}{2^2(1!)^2} + \frac{x^4}{2^4(2!)^2} - \frac{x^6}{2^6(3!)^2} + \frac{x^8}{2^8(4!)^2} - \cdots \tag{7.46}$$

Note that λ^2 is a constant for a given kinematic viscosity, fundamental frequency, and harmonic n.

We can also find a particular solution to the differential equation by setting the derivatives of $f_n(r)$ with respect to r equal solving for the constant C_3.

$$\frac{j^3\omega n}{\nu}C_3 = \frac{a_n}{\mu} \tag{7.47}$$

$$C_3 = \frac{a_n}{\mu}\frac{\mu}{\rho}\frac{1}{j^3\omega n} = \frac{-a_n}{j\rho\omega n} \tag{7.48}$$

The total solution becomes

$$f_n(r) = C_1 J_0(\lambda r) - \frac{a_n}{j\rho\omega n} \tag{7.49}$$

According to the no slip condition, velocity at the wall should be zero. That location is also defined by $r = R$, since $r = 0$ corresponds to the centerline of the vessel. In order to solve for the constant C_1 use the boundary condition of $u = 0$ at $r = R$.

$$0 = C_1 J_0(\lambda R) - \frac{a_n}{j\rho\omega n} \tag{7.50}$$

$$C_1 = \frac{a_n}{j\rho\omega n J_0(\lambda R)} \tag{7.51}$$

The total solution to the differential equation, for harmonic n, now becomes

$$f_n(r) = \frac{a_n}{j\rho\omega n}\left[\frac{J_0(\lambda r)}{J_0(\lambda R)} - 1\right] \tag{7.52}$$

This is still the complex form that is time independent. If we take the pressure gradient as the real part of $a_n e^{j\omega nt}$ and substitute u from Eq. (7.35), the corresponding velocity as a function of r and t becomes

$$u_n = \mathrm{Re}\left[\frac{a_n}{j\rho\omega n}\left\{\frac{J_0(\lambda r)}{J_0(\lambda R)} - 1\right\}e^{j\omega nt}\right] \tag{7.53}$$

Recall that this solution applies to the results of each harmonic. Now to find the velocity as a function of radius r and time t for the entire driving pressure, add together the results from all harmonics.

$$u(r) = \textstyle\sum_{n=1}^{\infty} u_n \tag{7.54}$$

Figure 7.3 shows an example plot of the total velocity $u(r)$ for two different r values in a mathematical model of an artery under pulsatile driving pressure. To get the flow resulting from these velocity profile,

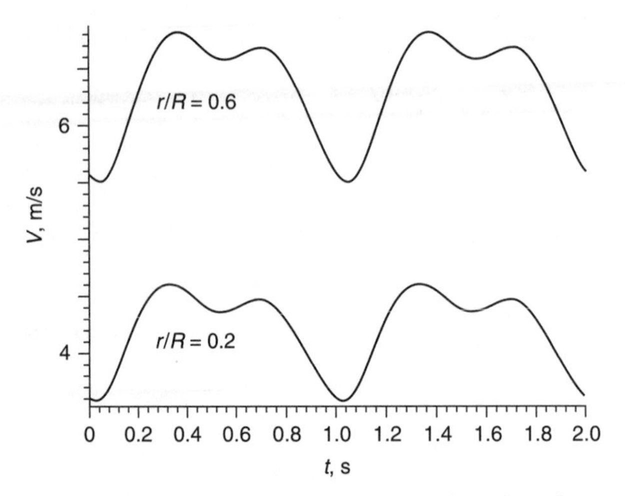

Figure 7.3 A graph of axial velocity as a function of radius and time, where r is the radius variable and R is the radius of the artery.

one needs only to integrate the velocities multiplied by the differential area, over the entire cross-section. The differential area as a function of r is $2\pi rdr$, so the flow term becomes

$$Q = \int_0^R u(r)\cdot 2\pi rdr \tag{7.55}$$

Wormersley (1955) published his solution when it was not so easy to make up a computer model of the flow. He integrated the velocity terms, solved for flow and published the following equation for the flow component resulting from each of the driving pressure gradient harmonics:

$$Q_n \frac{\pi R^4}{\mu} M_n \left(\frac{M_{10}}{\alpha^2}\right)_n \sin(\omega nt + \phi_n + \varepsilon_{10_n}) \tag{7.56}$$

The total flow Q, which can be compared to Q from Eq. (7.55), is

$$Q = \Sigma_{n=1}^{\infty} Q_n \tag{7.57}$$

where the pressure gradient associated with each harmonic is

$$\left.\frac{\partial P}{\partial x}\right|_n = M_n \cos(\omega nt + \phi_n) \tag{7.58}$$

The magnitude of the driving pressure is given by

$$M_n = \sqrt{A_n^2 + B_n^2} \text{ or alternatively, } M_n = |a_n|$$

The angle of the phase shift is given by

$$\phi_n = \tan^{-1}\left[\frac{B_n}{A_n}\right] \pm \overbrace{\frac{\pi}{2}}^{\text{depending on the quadrant}} \quad or\ \phi_n = \text{argument}\ (a_n)$$

Wormersley compiled the values of the constants for M_{10}/α_2 and for ε_{10} as a function of the alpha parameter. As alpha approaches zero, M_{10}/α_2 approaches 1/8 and ε_{10} approaches 90° and the solution becomes Poiseuille's law. Recall from Chap. 1 that α is the Wormersley number, or alpha parameter, which is a ratio of transient to viscous forces, and is defined by

$$\alpha = r\sqrt{\frac{\omega\rho}{\mu}}$$

Figure 7.4 shows a velocity waveform plotted in terms of time and vessel radius for a pulsatile flow condition in the uterine artery of a cow.

7.7 Pulsatile Flow in Rigid Tubes: Fry Solution

This solution was published by Greenfield and Fry (1965) for axisymmetric, uniform, fully developed, horizontal, newtonian, pulsatile flow and is comparable to the Wormersley solution in the sense that they are

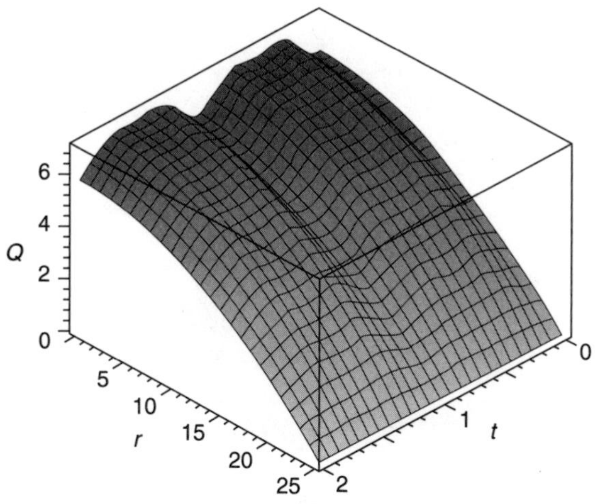

Figure 7.4 A three-dimensional plot that shows a flow waveform plotted in terms of time and vessel radius for a pulsatile flow condition. The curve was generated using Wormersley's version of the Navier-Stokes equation solution. The flow, Q, is shown in mm^3/s, the time t is shown in seconds, the radius r in millimeters.

both solutions relating flow rate to pressure gradient for the pulsatile flow condition. Since flow in arteries is pulsatile, this is an important case for human medicine. We will find the Fry solution particularly useful in Chap. 8 in modeling the behavior of an extravascular catheter/transducer pressure measuring system. This solution lends itself nicely to the development of an electrical analog to the pulsatile flow behavior.

In this solution one may begin from the Navier-Stokes equations where u is velocity in the x-direction (down the tube centerline), r is the radial direction variable (where $r = 0$ on the centerline), vr is radial velocity, v_Q is velocity in the transverse direction and ν is μ/ρ or the kinematic viscosity. Recall from Eq. (7.34) the differential equation for axisymmetic, uniform, fully developed horizontal, newtonian pulsatile flow.

$$\frac{a_n e^{j\omega nt}}{\rho} = \nu \frac{\partial^2(u)}{\partial r^2} + \frac{\nu}{r}\frac{\partial(u)}{\partial r} - \frac{\partial(u)}{\partial t}$$

Equation (7.34) can be rewritten in the form of Eq. (7.59).

$$\frac{\partial^2 u}{\partial r^2} + \frac{1}{r}\frac{\partial u}{\partial r} - \frac{1}{v}\frac{\partial u}{\partial t} = -\frac{1}{\mu}\frac{dP}{dx} \tag{7.59}$$

Two of the terms in Eq. (7.59) can be combined into

$$\frac{\partial^2 u}{\partial r^2} + \frac{1}{r}\frac{\partial u}{\partial r}$$

and be written from the chain rule as in Eq. (7.60).

$$\frac{1}{r}\left[(1)\frac{\partial u}{\partial r} + r\frac{\partial^2 u}{\partial r^2}\right] = \frac{1}{r}\frac{\partial\left(r\frac{\partial u}{\partial r}\right)}{\partial r} \tag{7.60}$$

Use the simplifying assumption that the pressure gradient

$$\frac{\partial P}{\partial x} = \frac{P_1 - P_2}{\ell}$$

where ℓ represents the length between points 1 and 2, and the governing differential equation from Eq. (7.59) can be rewritten:

$$\frac{1}{r}\frac{\partial\left(r\frac{\partial u}{\partial r}\right)}{\partial r} = \frac{1}{v}\frac{\partial u}{\partial t} - \frac{1}{\mu}\frac{(P_1 - P_2)}{\ell} \tag{7.61}$$

If we multiply both sides of Eq. (7.61) by $2\pi r dr$ and integrate from $r = 0$ to R where R is the tube radius, we get Eq. (7.62).

$$2\pi\int_0^R \frac{\partial\left(r\frac{\partial u}{\partial r}\right)}{\partial r}dr = \frac{1}{v}\int_0^R \frac{\partial u}{\partial t}(2\pi r)dr - \frac{(P_1 - P_2)}{\mu\ell}\int_0^R 2\pi r dr \tag{7.62}$$

The integral of derivative in the left-hand side of Eq. (7.62) returns the term $r(\partial u/\partial r)$ so we can rewrite Eq. (7.62) as shown in Eq. (7.63).

$$2\pi\, r \frac{\partial u}{\partial r}\bigg|_0^R = \frac{1}{v}\frac{\partial}{\partial t}\underbrace{\int_0^R u\,(2\pi r)dr}_{Q} - \frac{(P_1 - P_2)}{\mu \ell}\, 2\pi \frac{r^2}{2}\bigg|_0^R \tag{7.63}$$

Notice that the integral from zero to R of $u(2\pi r)dr$ is the integral of the velocity multiplied by the differential area so that the term is simply the flow through the tube. It is possible now to replace the first term on the right-hand side of the equation with the symbol Q, representing flow. After evaluating the last term between 0 and the vessel radius R we arrive at Eq. (7.64).

$$2\pi\, R \frac{\partial u}{\partial r}\bigg|_R = \frac{1}{v}\frac{\partial Q}{\partial t} - \frac{(P_1 - P_2)}{\mu \ell}\frac{2\pi R^2}{2} \tag{7.64}$$

By dividing both sides of Eq. (7.64) by πR^2 we arrive at Eq. (7.65).

$$\frac{2}{R}\frac{\partial u}{\partial r}\bigg|_R = \frac{1}{\pi R^2 v}\left(\frac{\partial Q}{\partial t}\right) - \frac{(P_1 - P_2)}{\mu \ell} \tag{7.65}$$

Recall that for a newtonian fluid, the shear stress at the wall is equal to the viscosity times the velocity gradient as shown in Eq. (7.66).

$$\tau_{\text{wall}} = -\,\mu \frac{\partial u}{\partial r}\bigg|_R \tag{7.66}$$

Substitute Eq. (7.66) into Eq. (7.65) to obtain Eq. (7.67)

$$\frac{2}{R}\left(\frac{-\tau}{\mu}\right) = \frac{\rho}{\pi R^2 \mu}\frac{\partial Q}{\partial t} - \frac{(P_1 - P_2)}{\mu \ell} \tag{7.67}$$

Next solve for the pressure gradient.

$$\frac{(P_1 - P_2)}{\ell} = \frac{\rho}{\pi R^2}\frac{\partial Q}{\partial t} + \frac{2\tau_w}{R} \tag{7.68}$$

If we solved the same equation for Poiseuille flow, we could use Eq. (7.69) to substitute for the shear stress at the wall. In that case the shear stress can be estimated from the flow rate as shown in Eq. (7.69).

$$\tau_w = \frac{8\mu Q}{2\pi R^3} \tag{7.69}$$

After the replacement of the shear stress term, Eq. (7.68) for the case of Poiseuille flow would be

$$\frac{(P_1 - P_2)}{\ell} = \frac{\rho}{\pi R^2}\frac{\partial Q}{\partial t} + \frac{2\dfrac{8\mu Q}{2\pi R^3}}{R} \tag{7.70}$$

$$\frac{(P_1 - P_2)}{\ell} = \frac{\rho}{\pi R^2}\frac{\partial Q}{\partial t} + \frac{8\mu Q}{\pi R^4} \tag{7.71}$$

Now we have an ordinary, first-order differential equation relating the flow rate and the time rate of change of flow to the pressure gradient. Recall from Eq. (7.69) that shear stress in Poiseuille flow depends on the vessel's hydraulic resistance and the flow rate. For the pulsatile flow case, assume that the shear stress depends on both the flow rate and the first derivative of the flow rate as shown by Eq. (7.72).

$$\frac{2\tau_w}{R} = R_v Q + L_I \frac{dQ}{dT} \tag{7.72}$$

The values for R_v and L_I in Eq. (7.72) are then given by Eqs. (7.68) and (7.74), respectively.

$$R_v = R_{\text{viscous}} = C_v \frac{8\mu}{\pi R^4} \tag{7.73}$$

$$L_I = L_{\text{inertia}} = \frac{c_1 \rho}{\pi R^2} \tag{7.74}$$

By substituting Eq. (7.72) into Eq. (7.78) we get:

$$\frac{P_1 - P_2}{\ell} = \frac{\rho}{\pi R^2}\frac{dQ}{dt} + \overbrace{R_v Q + L_I \frac{dQ}{dt}}^{eq\ 7.72} \tag{7.75}$$

Next, substituting the expressions for the viscous resistance term, R_v and the inertia term, L_I, yields the following equation. The two constants c_1 and c_v are constants, which may be solved for empirically or predicted by the Wormersley equation.

$$\frac{P_1 - P_2}{\ell} = \frac{\rho}{\pi R^2}\frac{dQ}{dt} + \frac{c_1 \rho}{\pi R^2}\frac{dQ}{dt} + \frac{c_v 8\mu}{\pi R^4} Q \tag{7.76}$$

$$\frac{P_1 - P_2}{\ell} = \frac{(1 + c_1)\rho}{\pi R^2}\frac{dQ}{dt} + \frac{c_v 8\mu}{\pi R^4} Q \tag{7.77}$$

Finally, it is possible to simplify the Fry solution to the following first order ordinary differential equation with terms that represent fluid inertance and fluid resistance as shown in Eq. (7.78).

$$\frac{P_1 - P_2}{\ell} = L\frac{dQ}{dt} + R_v Q \tag{7.78}$$

where

$$L = \frac{(1 + c_1)\rho}{\pi R^2} = \frac{c_u \rho}{\pi R^2} \tag{7.79}$$

$$c_u = 1 + c_1\left(\text{note that } L_I = \frac{c_1 \rho}{\pi R^2}\right) \tag{7.80}$$

$$R_v = \frac{c_v 8\mu}{\pi R^4} \tag{7.81}$$

It is possible to solve empirically for c_u and c_v or to predict those values from the Wormersley solution. For an alpha parameter of 7, $c_u = 1.1$ and $c_v = 1.6$.

7.8 Instability in Pulsatile Flow

It is helpful here to point out some useful information concerning stability and instability under pulsatile flow conditions. As a rule, flow in large arteries in humans is highly pulsatile. By the time the flow reaches small diameter arteries and arterioles the pulsatile elements drop out. By the time the flow reaches the capillaries, the flow becomes steady. Flow in large diameter veins becomes pulsatile once again.

The Reynolds number for flow in humans, even in large diameter arteries, is generally much less than 2000 and is therefore typically laminar. However, the presence of branching and other fluid wall interactions can sometimes result in local flow instabilities.

When there are disturbances throughout the flow field and throughout the entire oscillatory cycle, this condition constitutes turbulent flow. The presence of vortices at a specific location, or during a specific time in the pulsatile cycle, indicate a local instability only and not turbulent flow. For example, an instability associated with a valvular stenosis is an example of flow that is considered local flow instability as opposed to turbulent flow.

Bibliography

Munson BR, Young DF, Okiishi T. *Fundamental of Fluid Mechanics*. Wiley; 1994.

Greenfield JC, Fry DL. Relationship between instantaneous aortic flow and the pressure gradient. *Circ. Res.* October 1965; Vol. XVII:340–348.

Rainville ED, Bedient PE. *Elementary Differential Equations*, 5th ed. New York: Macmillan; 1974.

Del Toro V. *Principles of Electrical Engineering*, 2nd ed. Englewood Cliffs, NJ Prentice-Hall; 1972.

Wormersley JR. Method for the calculation of velocity, rate of flow and viscous drag in arteries when the pressure gradient is known. *J. Physiol.* 1955, 127, 553–563.

Chapter

8

Flow and Pressure Measurement

8.1 Introduction

From the measurement system for measuring hypertension in humans with an indirect method, to the system for measurement of flow and pressure waveforms in the uterine artery of cows, flow and pressure measurement systems have wide ranging applications in biology and medicine. A few of the varied methods for making these measurements will be discussed later.

Measurement of flow and pressure are two of the most important measurements in biological and medical applications. Because there is a complex control system in the human body, controlling pressure and flow, the interactions of the measuring system with the physiological control system increases the complexity of the measurement. The invasive nature of flow and pressure measurement also makes it difficult to obtain accurate measurements in some cases. Sometimes, a noninvasive or indirect measurement is possible.

According to the American Heart Association Council on High Blood Pressure Research, accurate measurement of blood pressure is critical to the diagnosis and management of individuals with hypertension. Pickering et al. (2005) wrote:

> The auscultatory (indirect measurement) technique with a trained observer and mercury sphygmomanometer continues to be the method of choice for measurement in the office.

8.2 Indirect Pressure Measurements

An important example of a noninvasive, indirect measurement is the measurement of blood pressure using a sphygmomanometer. This type measurement is routinely used in clinical practice because of its relative ease and low cost.

A sphygmomanometer has a cuff that can be wrapped around a patient's arm and inflated with air. The device includes a means for pressurizing the cuff, and a gauge for monitoring the pressure inside the cuff. When the pressure inside the cuff is increased above the patient's systolic pressure, the cuff squeezes the arm and blood flow through the vessels directly under the cuff is completely blocked. The pressure inside the cuff is released slowly, and when the cuff pressure falls below the patient's systolic pressure, blood can spurt through arteries in the arm beneath the cuff. The operator measuring blood pressure typically places a stethoscope distal to the cuff and just over the brachial artery. Turbulent flow of blood through the compressed artery makes audible sounds, known as Korotkoff sounds. At the time these Korotkoff sounds are initially detected, the blood pressure is noted, and this value becomes the measurement of the systolic pressure.

The nature of the sound heard through the stethoscope changes as the sphygmomanometer pressure continues to drop. The sounds change from a repetitive tapping sound to a muffled rumble and then the sound finally disappears. The pressure is once again noted at the instant that the sounds completely disappear. This value is the diastolic blood pressure.

Indirect measurement of blood pressure using a sphygmomanometer has the advantage of being noninvasive, inexpensive, and reliable. One disadvantage of the system is the relative inaccuracy of the system associated with operator error, the relatively low resolution, and the difficulty of measuring the diastolic pressure. Since the measurement of diastolic pressure depends on the detection of the absence of sound, the method is inherently dependent on the ability of the operator to clearly hear very quiet sounds and to judge the point in time at which those sounds disappear.

8.3 Direct Pressure Measurement

When more accurate pressure measurements are required, it becomes necessary to make a direct measurement of blood pressure. Two methods that will be described next, involve using either an intravascular catheter with a strain gauge measuring transducer on its tip, or the use of an extravascular transducer connected to the patient by a saline-filled tube.

8.3.1 Intravascular: strain gauge–tipped pressure transducer

Transducers are devices that convert energy from one form to another. It is often desirable to have electrical output represent a parameter like blood pressure or blood flow. Typically, these devices convert some type of mechanical energy to electrical energy. For example, most pressure measuring transducers convert stored energy in the form of pressure into electrical energy, perhaps measured as a voltage.

A pressure transducer very often uses a strain gauge to measure pressure. We can begin the topic of strain gauge tipped pressure transducers by considering strain. Figure 8.1 shows a member under a simple uniaxial load in order to demonstrate the concept of strain. As the load causes the member to stretch some amount ΔL, strain can be measured as $\Delta L/L$. A strain gauge could be bonded onto a diaphragm on a catheter tip.

Two advantages of a strain gauge pressure transducer, compared to an extravascular pressure transducer described in Sec. 8.3.2, are good frequency response, and fewer problems associated with blood clots. Some disadvantages might include the difficulty in sterilizing the transducers, the relatively higher expense, and transducer fragility.

As the strain gauge lengthens due to the load, the diameter of the wire in the strain gauge also decreases and therefore the resistance of the wire changes. Figure 8.2 shows a drawing of a strain gauge with a load F applied. It is possible to calculate the resistance of a wire based on its length, its cross-sectional area, and a material property known as resistivity. Equation 8.1 relates strain gauge resistance R to those properties.

$$R = \rho L/A \tag{8.1}$$

where R = resistance, in ohms, Ω
ρ = resistivity, in ohm $\times$ meters, Ωm
L = length, in meters, m
A = wire cross-sectional area, in square meters, m^2

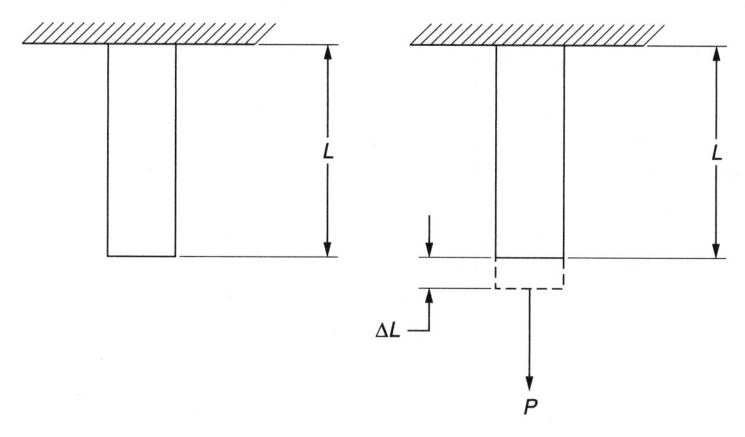

Figure 8.1 Demonstrating strain in a tensile specimen, $\Delta L/L$.

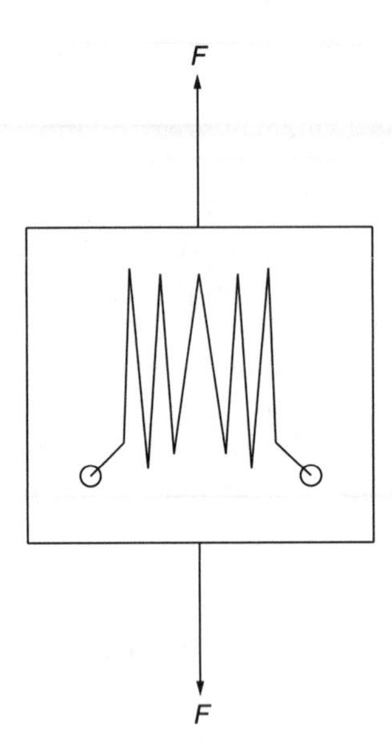

Figure 8.2 A strain gauge with force F applied.

We can now examine the small changes in resistance due to the changes in length and cross-sectional area by using the chain rule to take the derivative of both sides of Eq. (8.1). The result is shown in Eq. (8.2).

$$dR = \frac{\rho}{A}\,dL - \frac{\rho L}{A^2}\,dA + \frac{L}{A}\,d\rho \tag{8.2}$$

We can now divide Eq. (8.2) by $\rho L/A$ to get the more convenient form of the equation shown below in Eq. (8.3).

$$\frac{dR}{R} = \frac{dL}{L} - \frac{dA}{A} + \frac{d\rho}{\rho} \tag{8.3}$$

If you stretch a wire using an axial load, the diameter of the wire will also change. As the wire becomes longer, the diameter of the wire becomes smaller. Poisson's ratio is a material property which relates the change in diameter to the change in length of wire for a given material. Poisson's ratio is typically written as the character ν, which students should not confuse with kinematic viscosity, which is also typically written as ν. Poisson's ratio is defined in Eq. (8.4).

$$\text{Poisson's ratio } \nu = \frac{-dD/D}{dL/L} \tag{8.4}$$

In Eq. (8.4), D is the strain gauge wire diameter, and L is the strain gauge wire length.

It is also useful here to show that since $A = \pi/4\ D^2$, dA/A is related to dD/D so that we can write Eq. (8.3) in the more convenient terms of wire diameter instead of cross-sectional area.

$$\frac{dA}{A} = \frac{\frac{\pi}{4}(D_2^2 - D_1^2)}{\frac{\pi}{4}D^2} = \frac{\frac{\pi}{4}(D_2 - D_1)(D_2 + D_1)}{\frac{\pi}{4}D^2} = \frac{\frac{\pi}{4}(D_2 - D_1)(2D)}{\frac{\pi}{4}D^2} \tag{8.5}$$

In Eq. (8.5), dD is equal to $D_2 - D_1$ *and* $D_2 + D_1 = 2D$, where D is the average diameter between D_1 and D_2. Therefore, the relationship between dA and dD is shown by Eq. (8.6).

$$\frac{dA}{A} = \frac{2dD}{D} \tag{8.6}$$

Now it is possible to write an equation for the change in resistance divided by the nominal resistance as shown in Eq. (8.7),

$$\frac{dR}{R} = \frac{dL}{L} - \frac{2dD}{D} + \frac{d\rho}{\rho} \tag{8.7}$$

or by combining with the definition of Poisson's ratio we get Eq. (8.8).

$$\nu = \frac{-dD/D}{dL/L}$$

$$\frac{dR}{R} = \frac{dL}{L} + 2\nu\frac{dL}{L} + \frac{d\rho}{\rho} = (1 + 2\nu)\frac{dL}{L} + \frac{d\rho}{\rho} \tag{8.8}$$

The term $(1 + 2\nu)\ dL/L$ represents a change in resistance associated with dimensional change of the strain gauge wire. The second term $d\rho/\rho$ is the term that represents a change in resistance associated with piezoresistive effects, or change in crystal lattice structure within the material of the wire.

The gauge factor for a specific strain gauge is defined by Eq. (8.9).

$$\text{Gauge fa ctor } G = \left(\frac{\Delta R/R}{\Delta L/L}\right) = (1 + 2\nu) + \left(\frac{\rho/\Delta\rho}{\Delta L/L}\right) \tag{8.9}$$

For metals like nichrome, constantan, platinum-iridium, and nickel-copper the term $(1 + 2\nu)$ dominates the gauge factor. For semiconductor materials like silicon and germanium, the second term $\frac{\rho/\Delta\rho}{\Delta L/L}$ dominates the gauge factor.

A typical gauge factor for nichrome wire is approximately 2, and for platinum-iridium approximately 5.1. Strain gauges using semiconductor materials have a gauge factor that is two orders of magnitude greater than that of metal strain gauges. Semiconductor strain gauges are therefore more sensitive than metallic strain gauges.

If we know the gauge factor of a strain gauge and if we can measure $\Delta R/R$ then it is possible to calculate the strain, $\Delta L/L$. However, the change in resistance, ΔR is a very small change. Even though it is an electrical measurement, we need a strategy to detect such a small change in resistance.

A Wheatstone bridge is a device that is designed to measure very tiny changes in resistance. See the circuit diagram in Fig. 8.3

In Fig. 8.3, I_1 is the current flowing through the two resistors, R_1. The current flowing through the two resistors can be shown to be the excitation voltage driving the circuit divided by $2R_1$ as shown in Eq. (8.10).

$$I_1 = V_e/(2R_1) \tag{8.10}$$

The voltage at point 1, between the two resistors is

$$E_1 = I_1 \cdot R_1 = V_e/2 \tag{8.11}$$

The current I_2, flowing through the right side of the circuit, through the strain gauge and the potentiometer is

$$I_2 = V_e/(2R + \Delta R) \tag{8.12}$$

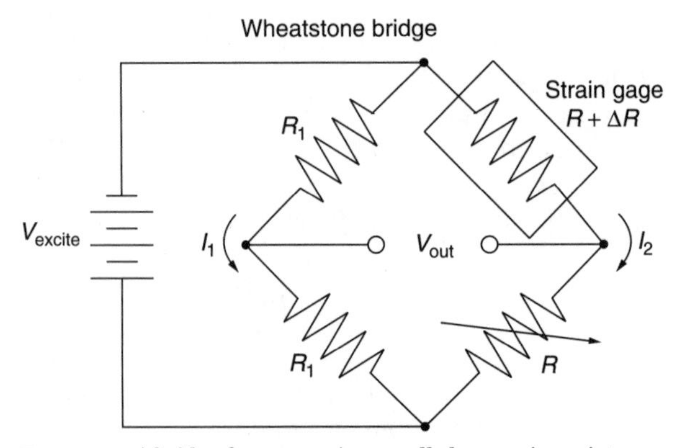

Figure 8.3 A bridge for measuring small changes in resistance.

The voltage at point two, on the right side of the circuit, between the strain gauge and potentiometer is

$$E_2 = I_2 \cdot R = V_e \times [R/(2R + \Delta R)] \tag{8.13}$$

V_{out} for the bridge is now given by the difference between the voltage at points 1 and 2. See Eqs. (8.14) through (8.16).

$$E_1 - E_2 = (V_e/2)\left\{1 - [2R/(2R + \Delta R)]\right\} \tag{8.14}$$

$$E_1 - E_2 = (V_e/2)\ [\Delta R/(2R + \Delta R)] \tag{8.15}$$

and since $\Delta R/R$ is much, much greater than $\Delta R/\Delta R$,

$$E_1 - E_2 = (V_e/4)(\Delta R/R) \tag{8.16}$$

Therefore, to measure $(\Delta R/R)$ we can use a bridge and measure $(V_{out}/V_{excitation})$. It is possible to obtain $\Delta R/R$ by multiplication of $(V_{out}/V_{excitation})$ by four. Now, if we divide $\Delta R/R$ by the gauge factor, G, we will obtain the strain.

If you question the necessity of the bridge in our measurement, compare the circuit in Fig. 8.4. If you use a known resistance in series with the strain gauge and measure resistance across the strain gauge, by measuring voltage, you will obtain the following:

The current I flowing through the circuit is

$$I = V_e/(2R + \Delta R) \tag{8.17}$$

The output voltage of the circuit that is used to measure the change in resistance is

$$V_{out} = IR = (V_e \cdot R)/(2R + \Delta R) \tag{8.18}$$

Since $R/2R >> R/\Delta R$, this circuit yields a fairly constant output of 1/2 V.

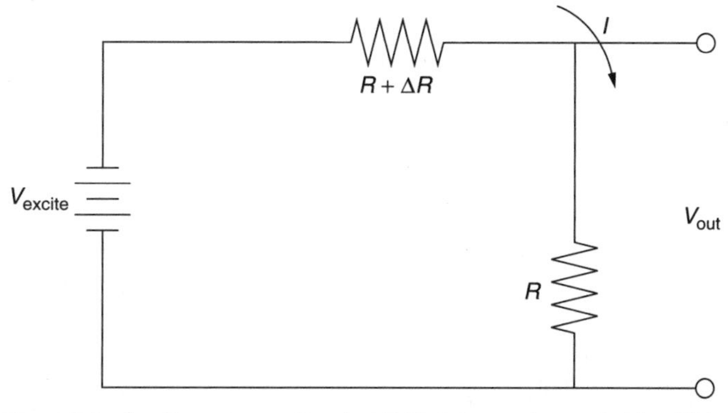

Figure 8.4 A voltage measuring circuit that does not use a bridge. This circuit can not accurately measure small changes in resistance.

Finally, pressure is proportional to the strain measured by the transducer and the strain gauge can be calibrated to output a voltage proportional to pressure. One can imagine a pressure transducer with a diaphragm that moves depending on the pressure of the fluid inside the transducer. A strain gauge mounted on the diaphragm of the transducer measures the displacement, which is proportional to pressure.

In a strain gauge-tipped pressure transducer, a very small strain gauge is mounted on the tip of a transducer, and the strain gauge tip deflects proportionally to the pressure measured by the strain gauge. The resultant strain may also be converted to a voltage output which may be calibrated to the pressure being measured.

8.3.2 Extravascular: catheter-transducer measuring system

Figure 8.5 shows a schematic of an extravascular pressure transducer. The transducer is connected to a long thin tube called a catheter. The catheter can transmit pressure from the blood vessel of interest to the extravascular pressure transducer. The pressure is transmitted through a column of heparinized saline. Heparin is used to prevent the blood from clotting and clogging the end of the catheter.

Although the extravascular pressure measurement system is a nice compromise between relatively low cost and accuracy, there are several potential problems associated with the system that I will mention here.

- Air bubbles in the system, when present, have an important effect on the system's ability to measure high frequency components of the pressure waveform. When you have air bubbles in the brake lines of your automobile, the brakes feel spongy and are less responsive. In

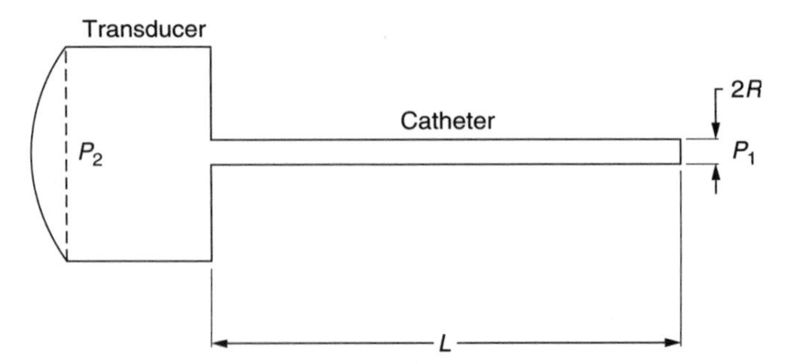

Figure 8.5 An extravascular pressure transducer connected to a long thin catheter which can be inserted into a blood vessel to transmit the pressure in that vessel.

the same fashion, an extravascular transducer with air bubbles in the catheter will be spongy and may not respond accurately to the pressure.

- Blood clots can form in the catheter. The blood clots will restrict flow and either plug the catheter or cause a significant pressure drop between the vessel and the extravascular transducer.
- The use of the extravascular pressure transducers require a surgical cut down. This disadvantage is also present in other direct pressure measurements but is not a disadvantage for indirect pressure measurements.

8.3.3 Electrical analog of the catheter measuring system

In Chap. 7, Sec. 7.7, a solution was developed that was published by Greenfield and Fry in 1965 that shows the relationship between flow and pressure for axisymmetric, uniform, fully developed, horizontal, Newtonian, pulsatile flow. The Fry solution is particularly useful when considering the characteristics of a transducer and catheter measuring system. By developing an electrical analog to a typical pressure measuring catheter, it will be possible for us to use some typical, well-known solutions to RLC circuits to characterize things like the natural frequency and dimensionless damping ratio of the system. From our circuit analog, we will be able to understand better the limitations of our pressure measuring system and to predict important characteristics.

It was possible to simplify the Fry solution to the following first order ordinary differential equation with terms that represent fluid inertance and fluid resistance as was shown next and in Chap. 7, Eq. (7.78).

$$\frac{P_1 - P_2}{\ell} = L\frac{dQ}{dt} + R_v Q$$

In Eq. (7.78), P_1 represents the pressure in the artery being measured, P_2 represents the pressure at the transducer, ℓ represents the length of the catheter, Q is the flow rate of the saline in the catheter, and dQ/dt is the time rate of change of the flow rate. Hydraulic inertance and hydraulic resistance are represented by L and R_v, respectively, and are defined in Eqs. (8.19) and (8.20).

$$L = \frac{(1 + c_1)\rho}{\pi R^2} = \frac{c_u \rho}{\pi R^2} \tag{8.19}$$

$$R_v = \frac{c_v\, 8\mu}{\pi\, R^4} \tag{8.20}$$

In Eq. (8.19), ρ represents the fluid (saline) density in the catheter and R represents the radius of the catheter. Three empirical proportionality constants are represented by c_u, c_1, and c_v. In Eq. (8.20), μ represents fluid viscosity (saline viscosity).

The volume compliance of the transducer, C, represents the stiffness of the transducer or the change in volume inside the transducer corresponding to a given pressure change. The volume compliance is written in Eq. (8.21).

$$C = \frac{dV}{dP_2} \tag{8.21}$$

By separating variables and integrating with respect to time, it is possible to solve for flow rate Q as a function of the change in the pressure in the transducer as shown in Eqs. (8.22) and (8.23).

$$dV = C\,dP_2 \tag{8.22}$$

$$Q = \frac{dV}{dt} = C\frac{dP_2}{dt} \tag{8.23}$$

The time rate of change of Q, as a function of compliance and P_2 can now be written:

$$\frac{dQ}{dt} = C\frac{d^2P_2}{dt^2} \tag{8.24}$$

Now it becomes possible to substitute Eqs. (8.23) and (8.24) into Eq. (7.88).

$$LC\frac{d^2P_2}{dt^2} + R_vC\frac{dP_2}{dt} = \frac{P_1 - P_2}{\ell} \tag{8.25}$$

L and R_v are defined by Eqs. (8.19) and (8.20), respectively. C is the volume compliance of the transducer. The length of the catheter is represented by ℓ, and the pressure in the blood vessel and the transducer are P_1 and P_2, respectively.

Next rewrite the derivative terms in the somewhat simplified form where dp/dt is written as $\dot{P}$ and d^2p/dt^2 as $\ddot{P}$ and we arrive at Eq. (8.26).

$$\ell LC\ddot{P}_2 + \ell R_vC\dot{P}_2 + P_2 = P_1 \tag{8.26}$$

Now we can define a term, E, that is equal to $1/C$ or the inverse of the volume compliance of the transducer. The term E is known as the volume modulus of elasticity of the transducer. Then, by multiplying both sides of the equation by the catheter length,ℓ, we arrive at Eq. (8.27). Notice that Eq. (8.27) is a second order, linear, ordinary differential equation with driving function EP_1. This type of equation is typical in many types of electrical and mechanical applications and one that we will use several times.

$$\underset{m}{LP_2} + \underset{b}{R_v P_2} + \underset{k}{EP_2} = EP_1 \tag{8.27}$$

For the mechanical analog system shown in Fig. 8.6, the equation defines a typical spring, mass, damper system where the spring constant is $k = E$, the damping coefficient for the damper is $b = \ell R_v$, and the mass term is $m = \ell L$. The canonical form of the second-order differential equation describing the spring, mass, damper system, which practically all mechanical engineers have seen, is shown in Eq. (8.28).

$$m\ddot{y} + b\dot{y} + ky = aF(t) \tag{8.28}$$

8.3.4 Characteristics for an extravascular pressure measuring system

For all second order systems, there are several system characteristics that may be of interest. In this section, we will discuss those characteristics, including static sensitivity, undamped natural frequency, and damping ratio.

For the mechanical system from Eq. (8.28), the characteristics are well known and are written in Eqs. (8.29) to (8.31).

In Eq. (8.29), y represents displacement of the mass shown in the mechanical spring mass and damper system in Fig. 8.6. The figure shows a time varying driving force $aF(t)$ driving the mass up and down while it is attached to a spring with spring rate k, and a damper with associated constant b.

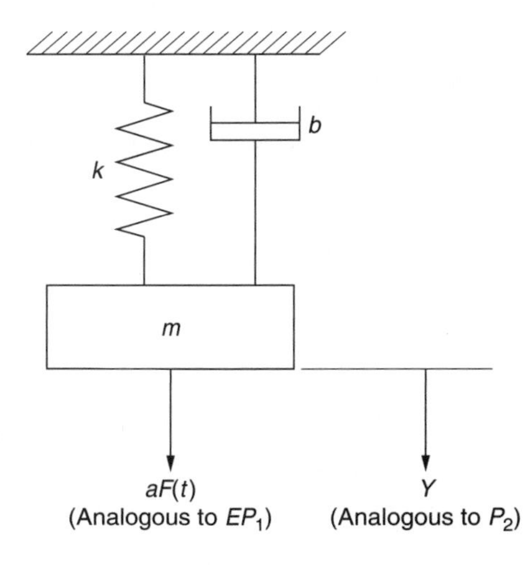

Figure 8.6 A spring, mass, damper analog to the extravascular pressure measuring system. The driving force for the system in the picture is $aF(t)$.

Static sensitivity

$$k = a/y \tag{8.29}$$

Undamped natural frequency

$$\omega_n = \sqrt{\frac{k}{m}}\,\frac{\text{rad}}{\text{s}} \tag{8.30}$$

Damping ratio

$$\zeta = \frac{b}{2\sqrt{km}} = \frac{b}{b_{cr}} \tag{8.31}$$

By making the appropriate substitutions from Eq. (8.27), we can now solve for the specific characteristics associated with our extravascular pressure measurement system.

$$\underset{m}{L} P_2 + \underset{b}{R_v} P_2 + \underset{k}{E} P_2 = EP_1$$

For the static system, displacement y does not change. In the pressure measuring system, the pressure does not change. All of the higher order terms like $\dot{y}$, $\ddot{y}$, $\dot{P}_2$, and $\ddot{P}_2$ are now zero. For the pressure measuring system, it is desirable that for every pressure input an equivalent pressure output occurs, so let us design the static gain to be unity. Equation 8.28 for the static system is

$$m\underbrace{\ddot{y}}_{0} + b\underbrace{\dot{y}}_{0} + ky = aF(t) \qquad \text{or} \qquad ky = aF(t)$$

For the analogous pressure measuring system $k = E$ and $a = E$ so the static sensitivity is 1, or E/E.

Again, this means simply that the measurement system is designed to measure the input value and repeat it as the output value.

The undamped natural frequency and dimensionless damping ratio of the system are given by Eqs. (8.32) and (8.33).

Undamped natural frequency

$$\omega_n = \sqrt{\frac{E}{\ell L}}\,\frac{\text{rad}}{\text{s}} = \sqrt{\frac{E\pi R^2}{\ell c_u \rho}}\,\frac{\text{rad}}{\text{s}} \tag{8.32}$$

Damping ratio

$$\zeta = \frac{\ell R_v}{2\sqrt{E\ell L}} = \frac{\ell \dfrac{c_v 8\mu}{\pi\, R^4}}{2\sqrt{E\ell \dfrac{c_u\rho}{\pi R^2}}} \tag{8.33}$$

Next I describe two examples of second order, pressure measuring systems. Case 1 is an undamped catheter measuring system, and Case 2 is the undriven, damped system.

8.3.5 Case 1—the undamped catheter measurement system

For the special case of a driven, undamped catheter measuring system we can now estimate some of the characteristics of the system. An undamped system would be one in which the viscosity of the fluid in the catheter, and therefore the corresponding hydraulic resistance in the catheter, is very small and therefore close to zero. A practical example of an undamped measuring system, is one in which the viscosity is very small compared to the mass of the system. A relatively large diameter catheter, with no air bubbles and with a relatively low viscosity fluid, like water, and relatively low total volume displacement would behave in this way. A system like this could have a very fast response time but would be susceptible to "catheter whip," or noise associated with accelerations at the catheter tip.

For this case recognize that viscosity μ, viscous resistance R_v, and the damping coefficient b are all approximately equal to zero. Then recall the expression for undamped natural frequency in Eq. (8.32). We can already see that the damping ratio is zero for this undamped system and we could predict the natural frequency from Eq. (8.32).

$$\mu \cong 0 \qquad R_v \cong 0 \qquad b \cong 0$$

Undamped natural frequency

$$\omega_n = \sqrt{\frac{E}{\ell L}}\,\frac{\text{rad}}{\text{s}} = \sqrt{\frac{E\pi R^2}{\ell c_u \rho}}\,\frac{\text{rad}}{\text{s}}$$

Damping ratio

$$\zeta = \frac{\ell R_v}{2\sqrt{E\ell L}} = 0$$

Going a step further, we would also like to predict the output of the pressure measuring system for every input. If P_1 is the input pressure to the system, then we would like the pressure output, P_2, to be equal to P_1 for any input frequency. Let us set the input, or driving pressure, to be $P_1\cos(\omega t)$. The output of the system or the pressure at the transducer is now $P_2\cos(\omega t)$. Recall the governing differential equation from Eq. (8.27).

$$\ell L\ddot{P}_2 + \ell R_v\dot{P}_2 + EP_2 = EP_1$$

Now after solving for the first and second derivative of P_2, the expressions below can be substituted into Eq. (8.27).

$$P_2 = P_2\cos(\omega t) \tag{8.34}$$

$$\dot{P}_2 = -\omega P_2\sin(\omega t) \tag{8.35}$$

$$\ddot{P}_2 = -\omega^2 P_2\cos(\omega t) \tag{8.36}$$

Equation 8.27 now becomes Eq. (8.37). Note that P_2 now represents a constant.

$$\ell L(-\omega^2)P_2\cos(\omega t) + \ell R_v(-\omega)P_2\sin(\omega t) + EP_2\cos(\omega t) = EP_1\cos(\omega t) \tag{8.37}$$

Since R_v is equal to zero for this case, the second term drops out and we can divide through by $\cos(\omega t)$. The result is Eq. (8.38).

$$\ell L(-\omega^2)P_2 + EP_2 = EP_1 \tag{8.38}$$

Solving for P_2 yields Eq. (8.39).

$$P_2 = \frac{EP_1}{(E - \ell L\omega^2)} = \frac{P_1}{\left(1 - \frac{\ell L\omega^2}{E}\right)} \tag{8.39}$$

Now using the definition of ω_n from Eq. (8.32) we can finally solve for the transfer function, P_2/P_1 as shown below in Eq. (8.41).

$$\frac{1}{\omega_n^2} = \frac{\ell L}{E} \tag{8.40}$$

$$\frac{P_2}{P_1} = \frac{1}{\left[1 - \frac{\omega^2}{\omega_n^2}\right]} \tag{8.41}$$

Finally, Fig. 8.7 shows the response of the second order, undamped, catheter pressure measuring system. Note that there is significant amplification of the signal as the input frequency approaches the natural frequency of the system.

8.3.6 Case 2—the undriven, damped catheter measurement system

A damped system would be one in which the viscosity of the fluid in the catheter, and therefore the corresponding hydraulic resistance in

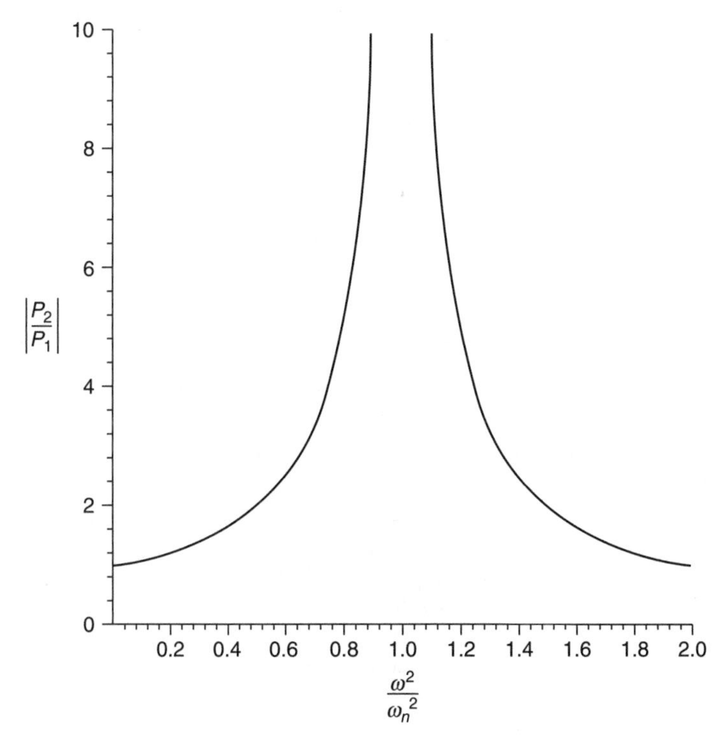

Figure 8.7 Response of a second order, undamped, pressure-measuring system.

the catheter, are relatively large and therefore significant. A practical example of a damped measuring system, is one in which the viscosity is relatively large compared to the mass of the system. A relatively small diameter catheter, with air bubbles or perhaps with a crimp in the tubing and with a relatively high viscosity fluid like blood, might behave in this way. A system like this could have a relatively low response time but might provide a kind of filtering action to eliminate higher frequency noise on the pressure waveform. In this context an undriven system is a system in which the input pressure may change, as with a step input, however, the system is not driven by an oscillating input pressure. Imagine the pop test described in Sec. 8.3.7 and shown in Fig. 8.8.

For the undriven system, we set the driving pressure equal to zero and Eq. (8.27) becomes Eq. (8.42).

$$\ell L\ddot{P}_2 + \ell R_v\dot{P}_2 + EP_2 = 0 \tag{8.42}$$

Since the system is undriven, we might imagine that when there is a step change in pressure, for example going from some positive pressure

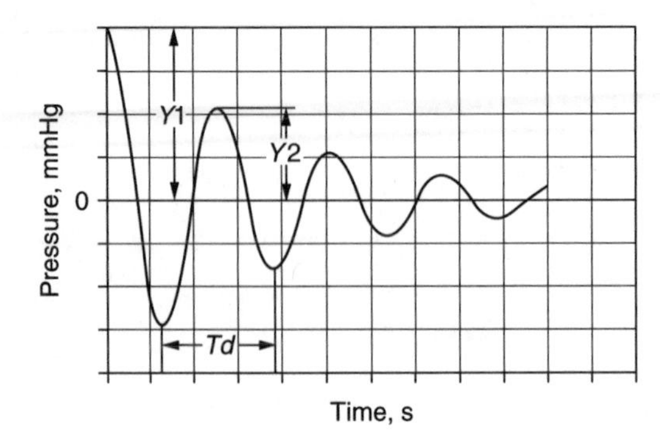

Figure 8.8 Step response for an under damped second order system.

to zero pressure, the system would follow an exponential decay. We can try $P_2 = Ae^{\lambda t}$, where λ is one divided by the time constant of the system. The first and second derivatives of P_2 now become $\dot{P}_2 = \lambda Ae^{\lambda t}$ and $\ddot{P}_2 = \lambda^2 Ae^{\lambda t}$. By substituting those values into Eq. (8.42) we arrive at Eq. (8.43).

$$\ell L\lambda^2 Ae^{\lambda t} + R_v \ell \lambda Ae^{\lambda t} + EAe^{\lambda t} = 0 \tag{8.43}$$

By using the substitutions suggested in Eq. (8.27), $\ell L = m$ and $R_v \ell = b$, we see a simplified version of Eq. (8.43).

$$m\lambda^2 + b\lambda + E = 0 \tag{8.44}$$

It is now possible to use the quadratic formula to solve for λ in terms of b, M, and volume modulus of elasticity of the transducer, E.

$$\lambda = \frac{-b \pm \sqrt{b^2 - 4mE}}{2m} \tag{8.45}$$

Recall from the definition of the damping ratio and critical damping, from Eq. (8.31), that the system is critically damped if it has a damping ratio of 1, which means that $b = b_{\text{critical}} = 2\sqrt{Em}$. By substituting that term into Eq. (8.45), it is possible to solve for the lambda associated with critical damping.

$$\lambda_{\text{critical}} = \frac{-b}{2m} \tag{8.46}$$

Recall that for the catheter measuring system, the general variables in Eq. (8.27) k, m, and b have the following definitions: $k = E$, $m = \ell L$, and $b = R_v \ell$.

$$\lambda_{\text{critical}} = \frac{\ell R_v}{2\sqrt{E\ell L}} \tag{8.47}$$

Note that because we set the damping ratio to 1, i.e., when $b = 2\sqrt{Em}$, then Eq. (8.47) yields the same expression for the damping ratio in Eq. (8.33). Again, λ is 1 over the time constant of the system and ζ is the dimensionless damping ratio.

$$\text{Damping ratio} \equiv \zeta = \frac{\ell R_v}{2\sqrt{E\ell L}}$$

The output of our system is therefore defined by the expression shown in Eq. (8.48),

$$P_2 = Ae^{\lambda t} + Bte^{\lambda t} = Ae^{-\frac{b}{2m}t} + Bte^{-\frac{b}{2m}t} \tag{8.48}$$

where $b = R_v\ell$ and $m = \ell L$. Substituting $b = 2\sqrt{Em}$ into Eq. (8.48), also yields

$$P_2 = Ae^{-\sqrt{(E/m)t}} + Bte^{-\sqrt{(E/m)t}} \tag{8.49}$$

Comparing the definition for the natural frequency from Eq. (8.32), we find that we now have an expression for critically damped catheters that relates the output of the measuring system to the natural frequency of the catheter.

$$P_2 = Ae^{-\omega_n t} + Bte^{-\omega_n t} \tag{8.50}$$

Critically damped systems are a very important case. This kind of system is on the borderline, but does not oscillate as it returns to equilibrium after perturbation. This disturbance is damped out as quickly as possible, as in the case of a door closer designed to close a screen door as quickly as possible with no back and forth motion.

The critical time constant for the system is $1/\lambda_{\text{critical}}$ as shown in Eq. (8.51).

$$\text{Critical time constant} = \frac{2m}{b} = \frac{2\ell L}{R_v\ell} = \frac{2L}{R_v} = \frac{c_u r_o^2}{c_v 4\nu} \tag{8.51}$$

It is useful to point out that the Eqs. (8.52) and (8.53) are general expressions for the catheter measuring system and not a special case for the critically damped system.

$$\zeta = \frac{b}{2\sqrt{EM}} = \frac{R_v\ell}{2\sqrt{E\ell L}} \tag{8.52}$$

$$\omega_n = \sqrt{\frac{k}{m}} = \sqrt{\frac{E}{\ell L}} \tag{8.53}$$

8.3.7 Pop test—measurement of transient step response

One way to determine some characteristics of a second order system is to create a step input to the system and measure the response. One straightforward and simple method of measuring the system response is to input a step change into the system by popping a balloon or surgical glove and measuring the response. Imagine a catheter with one end connected to a pressure transducer and the other end inserted into a container covered with a balloon or surgical glove. If the system is pressurized and the balloon or glove is popped, the resulting pressure output from the system looks like that shown in Fig. 8.8. The original steady state pressure was Y_1 and the pressure output of the transducer oscillates back and forth around zero, depending on the characteristics of the pressure measuring system.

The period of the oscillation is labeled as T_d in Fig. 8.8. The damped natural frequency of the system, measured in rad/s, is given by the equation $\omega_D = 2\pi/T_D$.

The logarithmic decrement is given by δ and is defined by the equation $\delta = \ln[P_2(0)/P_2(1)]$.

The damping ratio of the pressure measuring system can now be calculated from the logarithmic decrement as shown in Eq. (8.54).

$$\zeta = \sqrt{\left[\frac{\delta^2}{4(\pi)^2 + \delta^2}\right]} \tag{8.54}$$

The relationship between the natural frequency, the damped natural frequency, and the damping ratio is shown below in Eq. (8.55).

$$\frac{\omega_D}{\omega_n} = \sqrt{1 - \zeta^2} \tag{8.55}$$

By measuring the amplitude ratio of successive positive peaks, it is possible to determine the logarithmic decrement of the system. From the logarithmic decrement, it then becomes straight forward to measure the damped natural frequency, the damping ratio, and the undamped natural frequency.

If the natural frequency of the catheter measuring system is very low, some relatively low-frequency harmonics of the pressure waveform can be amplified, distorting the pressure waveform significantly. If possible, choose a catheter-transducer measuring system with a natural frequency that is 10 times greater than the highest harmonic frequency of interest in the pressure waveform.

8.4 Flow Measurement

Measurement of blood flow is an important indicator of the function of the heart and the cardiovascular system in general. Cardiac output is the measurement of the flow output of the heart and is a measure of the ability of the heart and lungs to provide oxygenated blood to tissue throughout the body. In this chapter we will have a look at several techniques for measuring cardiac output as well as techniques for measuring blood flow in specific vessels as a function of time.

Adolph Fick was born in 1829 in Kassel, Germany. Fick studied medicine, later became interested in physiology, and took a position in Zurich where he was already making contributions to the scientific literature in physics at the age of 26. Fick's first contribution as a physicist was a statement that diffusion is proportional to concentration gradient. Fick is most famous because of a brief, obscure publication in 1870, in which he described how mass balance might be used to measure cardiac output—the Fick principle.

8.4.1 Indicator dilution method

The general formulation of the indicator dilution method of measuring blood flow is one that is based on the Fick principle. The substance is injected into the blood stream in a known quantity. The flow rate of blood through a given artery, which is also the time rate of change of volume of blood moving through the same artery, can be predicted if the rate of injection of some mass of the indicator is known, along with the change in concentration of the indicator in the blood. For example, imagine that a bolus of indicator dye is injected into the blood stream. The concentration of the dye at a single point in some vessel could be measured and will be high immediately upon injection and will reduce with time as the dye is first diluted and then cleared from the system. The concentration will be a function of time. For this case change in concentration, ΔC, would be the change in concentration over time at a specific point.

The general formulation for the flow rate is given by Eq. (8.56), where m is the given quantity of indicator substance injected, dm/dt is the time rate of injection of m, V is the blood volume, and C is the concentration of m in blood. Depending on the specifics of the method used, ΔC may be a change in concentration at a single point in a blood vessel at two different times, or it may be the instantaneous difference in concentration between two different locations. Examples demonstrating the indicator-dilution method are described in more detail in the following sections.

$$\text{Flow} \equiv Q = \frac{dV}{dt} = \frac{dm/dt}{\Delta C} \tag{8.56}$$

8.4.2 Fick technique for measuring cardiac output

The Fick principle relates the cardiac output, Q, to oxygen consumption and to the arterial and venous concentration of oxygen as shown in Eq. (8.57).

$$Q = \frac{dV}{dt} = \frac{dm/dt}{\Delta C} = \frac{dm/dt}{C_a - C_v} \tag{8.57}$$

For Eq. (8.57), Q is the cardiac output in L/min, dm/dt is the consumption of O_2 in L/min, Ca is the arterial concentration of O_2 in L/L, and C_v is the venous concentration of O_2 in L/L. Using a device called a spirometer, it is possible to measure a patient's oxygen consumption. It is also possible to measure both arterial and venous oxygen concentration from a blood sample collected through a catheter.

Oxygen is the indicator used in this method which enters the blood stream through the pulmonary capillaries. Cardiac output is calculated based on the oxygen consumption rate and the oxygen concentration in arterial and venous blood.

8.4.3 Fick technique example

A patient's spirometer oxygen consumption is 250 mL/min while her arterial oxygen concentration is 0.2 mL/mL and her venous oxygen concentration 0.15 mL/mL. Cardiac output is calculated by dividing the oxygen consumption rate by the change in concentration, (0.22 − 0.15) L/L. The patient's cardiac output is 5 L/min.

$$Q = \frac{dV}{dt} = \frac{dm/dt}{\Delta C}$$

$$Q = \frac{dV}{dt} = \frac{dm/dt}{\Delta C} = \frac{0.25 \text{ L/min}}{0.20 \text{ L/L} - 0.15 \text{ L/L}} = 5\,\frac{\text{L}}{\text{min}}$$

8.4.4 Rapid injection indicator-dilution method—dye dilution technique

Indocyanine green is used clinically as an indicator to measure blood flow. This method could be used to measure cardiac output but is also applicable to local flow measurements like cerebral blood flow or femoral arterial flow. A curve is generated using a constant flow pump and blood is continuously drawn from the vessel of interest into a colorimeter cuvette to measure color (dye concentration).

Figure 8.9 shows a dye concentration curve that is generated by this method. A bolus of dye is rapidly injected into the vessel of interest. The solid line in the figure represents the fluctuation in concentration of the

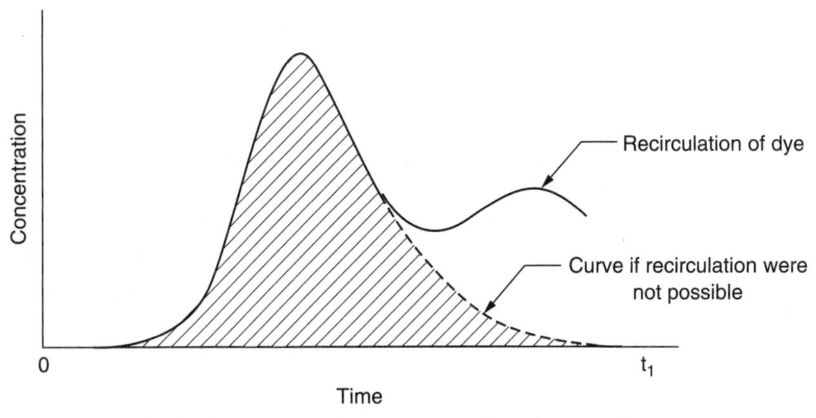

Figure 8.9 Recirculation curve used for measuring blood flow with indocyanine green dye dilution method.

dye, in the blood, after injection. The variable m represents the amount of dye injected. Q represents the flow rate and C represents the dye concentration. The area under the curve in Fig. 8.9 represents the amount of dye injected. The flow rate can be calculated as shown in Eq. (8.58).

$$Q = \frac{m}{\int_0^{t_1} C(t)dt} \tag{8.58}$$

In Eq. (8.58) Q is approximately constant and C is a function of time.

8.4.5 Thermodilution

The technique of thermodilution uses heat as an indicator for measuring blood flow, thereby avoiding the use of dyes like indocyanine green. Cooled saline can be injected into the right atrium while a thermistor placed in the pulmonary artery measures temperature. Temperature is used as a measure of the concentration of the indicator substance, in this case heat. Flow can then be calculated using Eq. (8.59).

$$Q = \frac{q}{\rho_b c_b \int_0^{t_1} \Delta T_b\,(t)dt} \frac{m^3}{S} \tag{8.59}$$

where Q = flow rate, $\mathrm{m^3/s}$
q = heat content of injectate, J
ρ_b = density of blood, $\mathrm{kg/m^3}$
c_b = specific heat of blood, J/(kg °K) = Nm/(kg °K)
ΔT = temperature change, °K.

8.4.6 Electromagnetic flowmeters

A conductor moving through a magnetic field generates an electromotive force (EMF) in that conductor. The EMF results in the flow of current which is proportional to the speed of the conductor. Based on this principle, an electromagnetic flowmeter can measure flow of a conducting fluid when that conducting fluid flows through a steady magnetic field. Figure 8.10 shows a schematic of a blood vessel between permanent magnets that generate a magnetic field within the blood vessel. A current is generated, between the two electrodes shown, that is proportional to the average blood velocity across the cross section of the vessel. The EMF generated between the electrodes placed on either side of the vessel is given in Eq. (8.60).

$$e = \int_0^{L_1} \vec{u} \times \vec{B} \cdot d\vec{L} \tag{8.60}$$

where $\vec{u}$ = the instantaneous blood velocity, m/s
$\vec{L}$ = the length between electrodes, m
$\vec{B}$ = the magnetic flux density, T
T has the units $\text{Wb/m}^2 = \dfrac{\text{V-s}}{\text{m}^2}$.

The variable e represents the EMF or voltage generated in response to blood velocity. For a uniform magnetic flux density B and a uniform velocity profile, the expression may be simplified to the scalar Eq. (8.61).

$$e = BLu \tag{8.61}$$

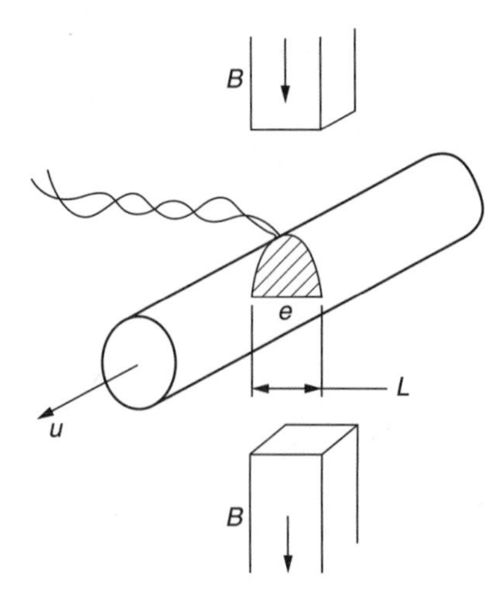

Figure 8.10 A schematic of an electromagnetic flowmeter with magnetic field $\vec{B}$ and blood velocity $\vec{u}$.

Some potential sources of error in electromagnetic flowmeters include:

1. Current flows from high to low velocity areas.
2. Current can be shunted through the vessel wall. This effect varies with hematocrit.
3. Fluid outside the vessel wall can shunt current.
4. Nonuniform magnetic flux density will cause a variation in the flowmeter output, even at a constant flow rate.

One typical type of electromagnetic probe is the toroidal type cuff perivascular probe. The toroidal cuff uses two oppositely wound coils and a permalloy core. The probe should fit snug around the vessel and this requires some constriction during systole.

8.4.7 Continuous wave ultrasonic flowmeters

Another type of flowmeter used in medical applications is the continuous wave Doppler flowmeter. This type of flowmeter is based on the Doppler principle, and when a target recedes from a fixed sound transmitter, the frequency is lowered because of the Doppler effect. For relatively small frequency changes, the relationship in Eq. (8.62) is true.

$$\frac{F_d}{F_o} = \frac{u}{c} \tag{8.62}$$

In Eq. (8.62), F_d is the Doppler shift frequency, F_o is the source frequency, u represents the target velocity in this case an erythrocyte, and c represents the velocity of sound. For two shifts, one from source and one to target:

$$\frac{F_d}{F_o} \approx \frac{2u}{c} \tag{8.63}$$

Consider now a probe at an angle α with respect to the patients skin, as shown in Fig. 8.11. The signal can now be calculated as:

$$F_d = \frac{2F_o u \cos\alpha}{c} \tag{8.64}$$

Some problems that occur with continuous wave Doppler ultrasound are:

1. The Doppler shifted frequency is actually not a single frequency, but a mixture of many frequencies.

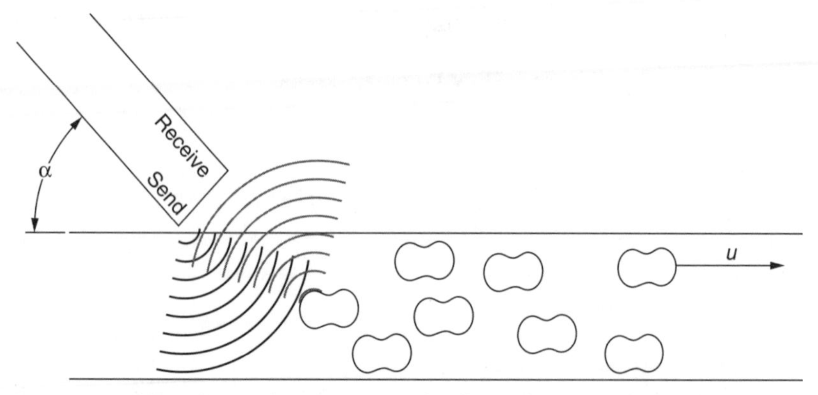

Figure 8.11 A Doppler continuous wave probe elevated at an angle α with respect to the patient's skin.

2. The velocity profile of the flow field is not constant across the vessel cross section.
3. The beam of the sound wave is broad and spreading.
4. Turbulence & tumbling of cells also causes a Doppler shift.
5. Simple meters do not measure reverse flow.

8.4.8 Continuous wave Doppler ultrasound example

Using a continuous wave Doppler with a carrier frequency of 7 MHz $\alpha = 45°$, blood velocity of 150 cm/s, and speed of sound = 1500 m/s, find the Doppler shifted frequency. Is it in the audible range?

$$F_d = \frac{2F_o u \cos\alpha}{c}$$

$$F_d = \frac{2(7 \times 10^6 \text{ Hz})(1.5 \text{ m/s})\cos(45)}{(1500 \text{ m/s})} \approx 10 \text{ kHz}$$

Yes. The frequency, 10 kHz is in the audible range of a normal, healthy human.

8.5 Summary and Clinical Applications

Chapter 8 introduces flow and pressure measurement schemes. A very quick review to remind the reader of the important clinical applications of those measurements is given below.

Indirect pressure measurements. According to the American Heart Association, high blood pressure is listed as a primary or contributing cause of death in more than a quarter of a million Americans each year.

Nearly one in three U.S. adults has high blood pressure and 30 percent of those people do not know it. Indirect blood pressure measurement using a sphygmomanometer is the method of choice for measuring blood pressure in the doctor's office and is critical to the management regimen of this disease.

Intravascular pressure measurement. Cardiologists sometimes insert a strain gauge tipped pressure catheter into an artery to determine whether a specific stenosis is the cause of decreased blood flow. If the pressure downstream from the blockage is much lower than the pressure immediately upstream, this indicates that the lesion is the cause. This procedure can also be used to evaluate the effectiveness of catheterization and stenting. This use of intravascular pressure measurement is an important aspect of the diagnosis of coronary arterial disease.

Extravascular pressure measurement. Pulmonary capillary wedge pressure is the pressure measured by inserting a balloon-tipped catheter from a peripheral vein into the right atrium, through the right ventricle and then positioning the catheter within a branch of the pulmonary artery. Measurement of pulmonary capillary wedge pressure (PCWP) gives an indirect measure of left atrial pressure and is particularly useful in the diagnosis of left ventricular failure and mitral valve disease. The catheter used for this procedure has one opening (port) at the tip of the catheter and a second port several centimeters proximal to the balloon. These ports are connected to extravascular pressure transducers allowing the pressure measurement.

Cardiac output measurement. Cardiac output is one of the main determinants of organ perfusion (Kothari et al., 2003). During a coronary artery bypass, dislocation of the heart changes the cardiac output and so it is important for the anesthesiologist to have a reliable tool for assessing hemodynamic status to avoid situations with disastrously low cardiac output. There are a number of methods used to assess cardiac output, including the Fick technique, rapid injection dye dilution and thermodilution.

Electromagnetic and continuous wave Doppler flowmeters. These two devices are two tools in the physicians' arsenal that can be used to measure blood flow in a variety of arteries. For example, physicians use Doppler flowmeters to diagnosis peripheral arterial disease. The physician measures flow waveforms and could compare, for example, the waveforms from arteries in the left arm to those from the right arm. A significant difference in the waveforms between the two vessels could indicate a stenosis in one of the vessels.

While electromagnetic flowmeters are more invasive, they are commonly used as the gold standard for measuring blood flow in animal experiments in laboratory settings. In Chap. 10, a model of flow through the mitral valve in the porcine model is presented (Szabo et al., 2004). That model was developed from data that was collected at the University of Heidelberg and the flow measurements were made using the electromagnetic flowmeter.

Bibliography

John WC Jr., Michael RN, Walter HO, et al. *Medical Instrumentation Application and Design*. John G. Webster, ed. 3rd ed. NY. Houghton Mifflin Company.

Cromwell L, Weibell FJ, Pfeiffer EA. *Biomedical Instrumentation and Measurements*. 2nd ed. Prentice-Hall. Englewood Cliffs, NJ; 1980.

Togawa T, Tamura T, Öberg P. *Biomedical Transducers and Instruments*. Boca Raton, FL. CRC Press; 1997.

Khandpur RS. *Biomedical Instrumentation, Technology and Applications*. NY. McGraw-Hill; 2005.

Pickering TG, Hall JE, Appel LJ, Falkner BE, Graves J, Hill MN, Jones DW, Kurtz T, Sheps SG, Roccella EJ. Recommendations for Blood Pressure Measurement in Humans and Experimental Animals, *Hypertension*. 2005; 45:142.

Kothari N, Amaria T, Hegde A, Mandke A, Mandke NV. Measurement of cardiac output: Comparison of four different methods, *IJTCVS*. 2003; 19:163–168.

Szabo G, Soans D, Graf A, Beller C, Waite L, Hagl S. A new computer model of mitral valve hemodynamics during ventricular filling, Euro. J. Cardiothorac Surg. 2004; 26:239–247.

Chapter

9

Modeling

9.1 Introduction

Models are used widely in all types of engineering, and especially in fluid mechanics. The term model has many uses, but in the engineering context, it usually involves a representation of a physical system, a prototype, that may be used to predict the behavior of the system in some desired respect. These models can include physical models that appear physically similar to the prototype or mathematical models, which help to predict the behavior of the system, but do not have a similar physical embodiment. In this chapter, we will develop procedures for designing models in a way to ensure that the model and prototype will behave similarly.

Consider the question, "Why can a 1.5-mm-long flea fall a meter without injury but a 1.5-m-tall man can't fall 1 km unhurt?" A model is a representation of a physical system that may be used to predict the behavior of the system in some desired respect. The implied question in the stated question above is, "Can the flea be used to model the man?"

It is true that a 1.5-mm-long flea can jump about 1 m. If we use geometric scaling, we might predict that a 1.5-m-long person would jump about 1 km. Since that is clearly not true, then what is wrong with this picture? A careful look at the theory of models will help answer our question.

In Sec. 9.7, we attempt to use a flea to model a man (or vice versa) to model the terminal velocity of the prototype using the model. We will learn, by using the theory of models and dimensional analysis, why this model does not work very well.

9.2 Theory of Models

The theory of models is developed from the principles of dimensional analysis. These principles tell us that any prototype can be described by a series of dimensionless parameters, which I will call Pi terms in this chapter. For the prototype system, the parameter that we would like to measure, Π_1, can be represented as a function Φ of a set of n dimensionless Pi terms.

$$\Pi_1 = \Phi(\Pi_2, \Pi_3, \Pi_4, \ldots, \Pi_n) \tag{9.1}$$

If the relationship Φ between Pi terms describes the behavior of the system, then it would also be possible to develop a similar set of Pi terms for a model that has the same dimensional relationships.

$$\Pi_{1m} = \Phi(\Pi_{2m}, \Pi_{3m}, \Pi_{4m}, \ldots, \Pi_{nm}) \tag{9.2}$$

9.2.1 Dimensional analysis and the Buckingham Pi theorem

Edgar Buckingham (1867–1940) was an American physicist who first generated interest in the idea of dimensional analysis. In 1914, Buckingham published the article, "On Physically Similar Systems: Illustration of the Use of Dimensional Equations," *Phys Rev.* 1914;4:345–376. The Buckingham Pi theorem states:

> The number of independent dimensionless quantities required to describe a phenomenon involving k variables is n, where $n = k - r$, and where r is the number of basic dimensions required to describe the variable.

To demonstrate the use of the Buckingham Pi theorem, imagine that we would like to perform a test that describes the pressure drop per length of pipe as a function of other variables that affect the pressure gradient (Fig. 9.1).

The first step in the process is to choose three fundamental dimensions, which describe mechanical properties. One might choose from a list of fundamental dimensions like force, length, time, mass, temperature, charge, voltage, and the list would be extensive. In fact, it is pretty clear that charge, voltage, and temperature are not important in this problem, so we will choose force, length, and time. We will designate those dimensions by the characters F, L, and T representing force, length, and time, respectively.

The second and arguably most difficult step in the modeling process will be to determine all important variables. In some models, it will not be obvious which variables are important. Occasionally, to simplify a

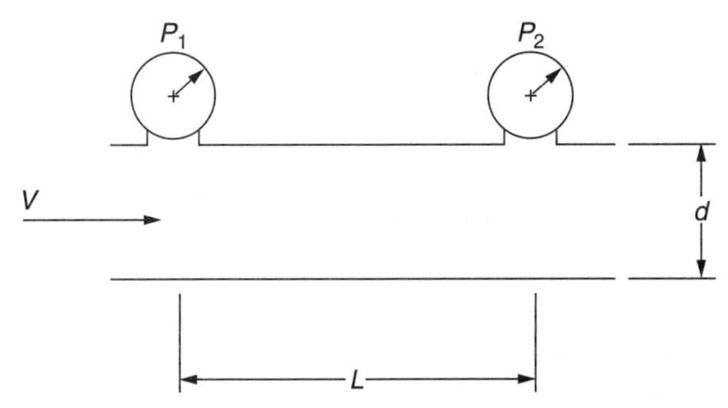

Figure 9.1 Pressure gauges at two points along a pipeline of diameter d, with a mean velocity of the pipe of V.

model, we will begin with a relatively small set of variables and when the model does not suitably represent the prototype we may find out that we need to add variables. For this example problem, we will assume that the pressure gradient $\Delta p/L$ is related to average velocity in the pipe, V, the diameter of the pipe, d, the viscosity of the fluid flowing in the pipe, μ, and the density of the fluid, ρ.

$$\Delta p/L = f(V, d, \mu, \rho) \tag{9.3}$$

The third step in the process is to write the dimensions of each variable. For example, the variable $\Delta P/L$ has the dimensions of (force/length^3) as shown next.

$$\Delta p/L \sim F/L^3 \tag{9.4}$$

The other variables in the example have dimensions as shown in Eqs. (9.5) through (9.7):

$$V \sim L/T \tag{9.5}$$

$$\mu \sim FT/L^2 \tag{9.6}$$

$$\rho \sim M/L^3 = FT^2/L^4 \tag{9.7}$$

Now that all of the dimensions have been described, we can use Buckingham's Pi theorem to determine the number of required Pi terms to describe this model.

$$n = k - r = 5 \text{ dimensions} - 3 \text{ basic dimensions} = 2 \text{ Pi terms} \tag{9.8}$$

The result of this analysis is that we have confirmed the following fact. To conduct an experiment to determine the relationship between $\Delta P/L$

and the other important variables, we only need to measure two dimensionless Pi terms and not five different variables.

Pi terms are dimensionless terms and in this case, it would be possible to choose the following two Pi terms to describe the system.

$$\Pi_1 = (\Delta p/L)d/(\rho V^2)$$

and

$$\Pi_2 = \rho V d/\mu$$

In fact, any two independent Pi terms would suffice. The inverse of either of the above Pi terms would also be valid Pi terms, for example, $1/\Pi_1$ or $1/\Pi_2$. The most important criteria of the two Pi terms is that they are independent terms. Another important factor in choosing the best Pi terms for the problem being studied, is that it is important to choose one Pi term that includes the dependent variable to be investigated and that it appears only in that term.

By using dimensional analysis, we have now learned that to run this experiment, we do not need to vary density, diameter, or viscosity. It is possible to vary only velocity and measure only pressure gradient. Not only have we reduced the variables from five to two, but we have also created dimensionless terms so that our results are independent of the set of units that we choose. Figure 9.2 shows a series of experiments that compare the pressure gradient, designated ΔP_ℓ in these charts, to the four variables V, d, ρ, and μ. These four comparisons require four separate graphs. This means that we will be able to generate the same amount of data from one experiment that we would have needed four experiments to generate, had we not used dimensional analysis.

Figure 9.3 shows a single dimensionless plot that contains the same information included in all four plots in Fig. 9.2. The plot shows the pressure gradient as a function of V, d, ρ, and μ.

9.2.2 Synthesizing Pi terms

In some models, Pi terms are relatively easy to determine. It is useful here to discuss an algorithm for generating independent Pi terms from a list of variables. To generate a set of independent Pi terms for this example problem, begin by writing down all of the variables, each raised to the power of an unknown as shown below.

$$(\Delta P/L)^{a} \quad (V)^{b} \quad (d)^{c} \quad (\mu)^{e} \quad (\rho)^{f}$$

Next, write the dimensions of those variables, in the same format.

$$(F/L^3)^{a} \quad (L/T)^{b} \quad (L)^{c} \quad (FT/L^2)^{e} \quad (FT^2/L^4)^{f}$$

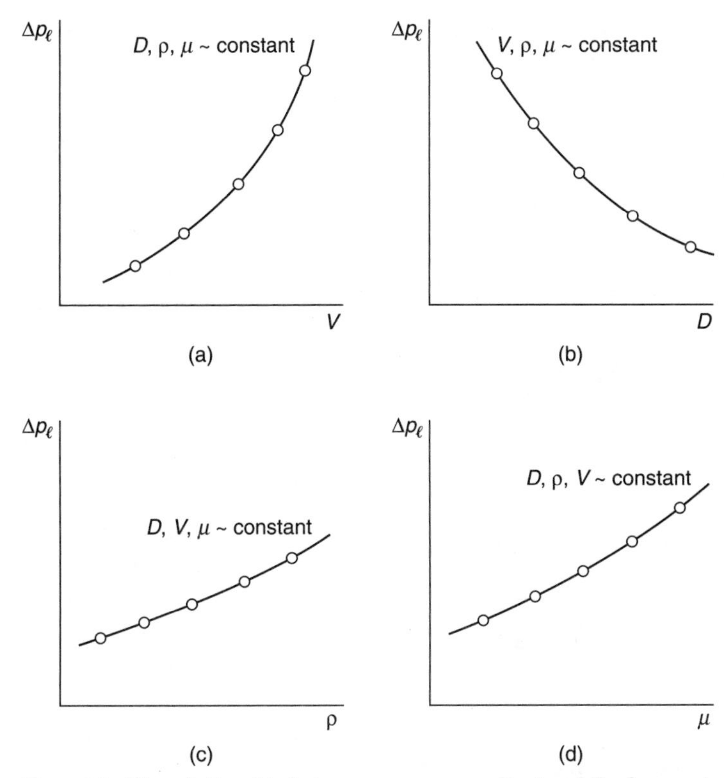

Figure 9.2 The relationship between pressure gradient and the four variables, velocity, pipe diameter, fluid density, and fluid viscosity. (Reprinted with permission of John Wiley and Sons Inc.)

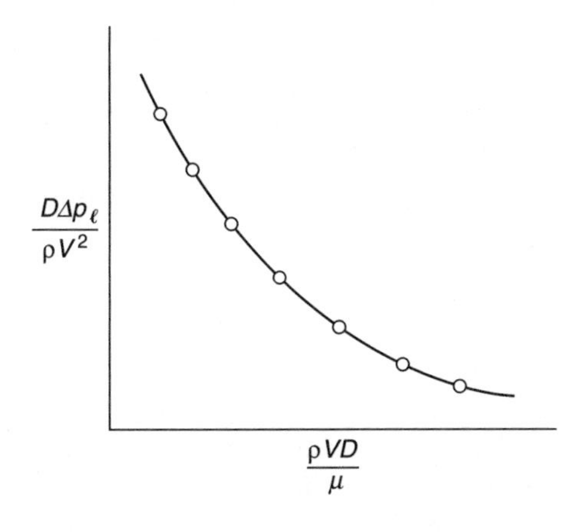

Figure 9.3 A single dimensionless plot that also includes all of the information included in the four plots in Fig. 9.2. (Reprinted with permission of John Wiley and Sons Inc.)

Now it is possible to write a set of three equations, one for each basic dimension, and to solve them simultaneously for three values. Since we have five unknowns and only three equations, it is necessary to assume the values of two convenient variables.

$$
\begin{aligned}
&\text{F:} \quad && a + 0 + 0 + e + f = 0 \\
&\text{L:} \quad && -3a + b + c - 2e - 4f = 0 \\
&\text{T:} \quad && 0 - b - 0 + e + 2f = 0
\end{aligned}
$$

For our example, we will assume $a = 1$ and $e = 0$ to produce two independent Pi terms. That will insure that $\Delta P/L$ will appear in only one Pi term. Solving the three equations yields the following:

$$
\begin{aligned}
&1 + f = 0 && f = -1 \\
&-3 + b + c - 4f = 0 && c = 1 \\
&-b + 2f = 0 && b = -2
\end{aligned}
$$

Therefore, by using the values of $a = 1$ and $e = 0$, one possible Pi term follows:

$$\Pi_1 = (\Delta P/L)^1 \, V^{-2} \, d^1 \, \rho - 1$$

To generate the second independent Pi term let $e = 1$ and $a = 0$. This will ensure that every variable appears in one of the Pi terms and that $\Delta P/L$ does not appear in Π_2.

$$
\begin{aligned}
&1 + f = 0 && f = -1 \\
&b + c - 2 - 4f = 0 && c = -1 \\
&-b + 1 + 2f = 0 && b = -1
\end{aligned}
$$

$$\Pi_2 = V^{-1} \, d^{-1} \, \mu \, \rho^{-1} = \mu/\rho V d$$

Note that Π_2 is the 1/Reynolds number, a well-known dimensionless parameter used in many fluid mechanics applications.

Therefore, if we make Π_2 for the model equal to Π_2 for the prototype, then Π_1 for the model will predict meaningful values for Π_1 of the prototype.

9.3 Geometric Similarity

When we equate Pi terms involving length ratios only, the model satisfies the condition of geometric similarity. Scale models in which all three dimensions of the model (height, width, and length) have the same scale have geometric similarity. A common dimensionless Pi term involving geometry is the ratio of length to width or the ratio of two lengths.

$$L_1/L_2 = L_{1m}/L_{2m}$$

If a model is a 1/10 scale model, this means that the ratio of lengths between the model and the prototype is 1/10. The flea in our example introduced above might be considered a 1/1000 scale model of a human, if it were similarly proportioned to the human. If we only know the flea's length and the man's height, we can consider the flea geometrically similar. We will call the flea a 1/1000 scale model for a first approximation.

But if we consider the overall effectiveness of the model, what about the other Pi terms that have forces and time in them? For a complete model, we also need some other types of similarity.

9.4 Dynamic Similarity

For a perfect model, we must be sure that we have included all important variables. Further, all Pi terms in the model must be equal to each Pi terms in the prototype. When we achieve geometric similarity by matching the geometric Pi terms, as with a scale model airplane, the model will look like a smaller version of the prototype but will not necessarily behave like the prototype. When we equate Pi terms involving force ratios we achieve dynamic similarity.

For example, density is a variable that contains force as one of its dimensions. A Pi term in some fluid mechanics problems that includes drag force and density could be the term that follows. For this Pi term the variable D represent drag force, ρ represents fluid density, V represents fluid velocity, and t_1 and t_2 are geometric dimensions.

$$\Pi_1 = [D/(\rho V^2 t_1 t_2)]$$

To achieve dynamics similarity, Π_1 in the prototype must equal Π_1 in the model.

$$[D/(\rho V_{\text{max}}^2\, t_1 t_2)]_p = [D/(\rho V_{\text{max}}^2\, t_1 t_2)]m$$

9.5 Kinematic Similarity

When we equate Pi terms involving velocity and/or acceleration ratios we obtain kinematic similarity. Many Pi terms that have a force dimension will also have a velocity or acceleration dimension. For example, the Reynolds number is a common Pi term in fluid mechanics that has both force dimensions and velocity dimensions. The force dimension shows up in both density and viscosity. To achieve kinematic similarity, and to achieve dynamics similarity, the model and the prototype must have Reynolds number similarity. That is, the Reynolds number of the model must equal the Reynolds number of the prototype.

$$\rho VD/\mu = (\rho VD/\mu)_m$$

To achieve similitude, we must have a model with geometric similarity, dynamic similarity, and kinematic similarity when compared to our prototype.

9.6 Common Dimensionless Parameters in Fluid Mechanics

Dimensionless Pi terms are used in many fluid mechanics applications. Some of the terms are used relatively frequently and have names. Some important dimensionless groups in fluid mechanics are shown in Table 9.1. In the table below ρ is density, V is fluid velocity, and L is some geometric length. Viscosity is represented by μ and gravitational acceleration is represented by g. Pressure is denoted by p. The frequency of the changing flow is ω. Surface tension in the Weber number is represented by σ.

9.7 Modeling Example 1—Does the Flea Model the Man?

If we use a flea to model a man (or vice versa) we need to decide on a few important parameters. Let us first assume that we want to model the terminal velocity of the prototype using the model. The terminal

TABLE 9.1 Shows Some Common Dimensionless Parameters

Pi term	Name	Ratio
$\rho VL/\mu$	Reynolds number, Re	Inertia/viscous
$V/(gL)^{1/2}$	Froude number, Fr	Inertia/gravity
$p/(\rho V^2)$	Euler number, Eu	Pressure/inertia
V/c	Mach number, Ma	Velocity ratio
$L(\omega\rho/\mu)^{1/2}$	Wormersley number, α	Inertia/surface tension
$\rho V^2L/\sigma$	Weber number, We	Inertia/surface tension

velocity is the maximum velocity that a falling object in free fall will reach. We could assume that the maximum velocity of the prototype will be a function of drag, D; the weight of the object, W; the geometry of the object (which we will designate by three dimensions L, t_1, and t_2); the density of the fluid in which the object is falling, ρ; and the viscosity of the fluid in which the object is falling, μ. Mathematically, we can write the relationship as follows:

$$V_{max} = f\,(D, W, L, t_1, t_2, \rho, \mu)$$

Therefore, for this modeling problem, there are eight variables and three basic dimensions that are required to describe the units on those variables. The three basic dimensions are time, length, and force. If our assumptions are valid, then the required number of Pi terms is $n = 5$.

$$n = k - r = 8 - 3$$

I will choose the following five independent Pi terms:

$\Pi_1 = \rho V_{max} L/\mu$	Reynolds number
$\Pi_2 = D/(\rho V_{max}^2 t_1 t_2)$	coefficient of drag
$\Pi_3 = D/W$	force ratio
$\Pi_4 = t_1/L$	geometrical scale
$\Pi_5 = t_2/L$	geometrical scale

For a first approximation for our model, we will take the flea weight to be 10^{-4} g. The flea length will be 1.5 mm and we will consider the flea and the man to have the same aspect ratio. That means that we will assume Π_4 and Π_5 to be identical in the model and prototype a priori. The man's length will be 1.5 m and the weight will be 100 kg.

Starting from this data, our geometric scale is 1000:1. The man is 1000 times longer and 1000 times wider than the flea. Because we are looking to model the terminal velocity of the falling man and/or the flea, the drag force will be equal to weight for $\Pi_3 = 1$.

For kinematic and dynamic similarity let $\Pi_{1_{man}} = \Pi_{1_{flea}}$ and let $\Pi_{2_{man}} = \Pi_{2_{flea}}$. The resulting equations are shown below.

$$\Pi_2 = (W/(\rho V_{max}^2 t_1 t_2))\text{man} = (W/(\rho V_{max}^2 t_1 t_2))_{flea}$$

$$(V_{man}/V_{flea})^2 = (W_{man}/W_{flea})(t_1 \times t_2)_{flea}/(t_1 \times t_2)_{man}$$

$$(V_{man}/V_{flea})^2 = (10^5/10^{-4})(1/1000)^2 = 1000$$

If the second Pi term is identical in the flea and the man, the terminal velocity of the man will be ~30 times that of the flea.

$$\pi_1 = (\rho V_{max} L/\mu)_{man} = (\rho V_{max} L/\mu)_{flea}$$

$$(V_{man}/V_{flea}) = L_{flea}/L_{man} = 1/1000$$

If the first Pi term, the Reynolds number, is identical in the flea and the man, then the velocity of the man will be 1000 times that of the flea. We can see that these two conditions are mutually exclusive and the man and the flea, falling through the same fluid are not a good model for each other. For Reynolds similarity, the man must fall 1000 times slower. For drag similarity, the man needs to fall 30 times faster.

If you want to allow the man and flea to fall at the same velocity and have Reynolds number similarity, it would also be theoretically possible to change fluids so that we have a different kinematic viscosity, ν. We might ask ourselves whether it is practical. The kinematic viscosity of standard air is $\nu_{air} = 1.46 \times 10^{-5}$ m^2/s. Now if we set the velocity of the man and the flea equal, with the constraint of maintaining Reynolds number similarity, the equation is shown below.

$$\frac{\left(\frac{\rho}{\mu}\right)_{man}}{\left(\frac{\rho}{\mu}\right)_{flea}} = \frac{l_{flea}}{l_{man}} = \frac{1}{1000}$$

Since viscosity divided by density is equivalent to kinematic viscosity; $\mu/\rho = \nu$, and if the flea is falling through air and the man is falling through an alternate fluid, it is also possible to write the equation as:

$$\frac{\nu_{air}}{\nu_{alternate}} = \frac{1}{1000}$$

For Reynolds number similarity the man needs to fall through a fluid that has a kinematic viscosity 1000 times greater than that of standard air or ~1.5×10^{-2} m^2/s. Glycerine at 20 °C has a kinematic viscosity of 1.2×10^{-3} m^2/s, so we would need to use glycerine at a much cooler temperature to achieve that kinematic viscosity! Because of glycerine's very large density compared to air, now buoyancy would become a significant factor and we realize that the flea just is not a good model for the man (and vice versa).

Our original question was, "Why can a 1.5 mm flea fall a meter without injury but a 1.5 m man can't fall 1 km unhurt?" The answer is that the model and prototype are geometrically similar (1/1000 scale) but not dynamically and kinematically similar. The low Reynolds number and relatively high drag effect predict that the flea falls very slowly!

9.8 Modeling Example 2

We would like to study the flow through a 5 mm diameter venous valve carrying blood at a flow rate of 120 mL/min. We will use water instead of blood, which is more difficult to obtain and more difficult to work with. Take the viscosity of blood to be 0.004 Ns/m^2 and the viscosity of water to be 0.001 Ns/m^2. Complete geometric similarity exists between the model and prototype. Assume a model inlet of 5 cm in diameter. Determine the required flow velocity in the model that would be required for Reynolds number similarity.

Begin by calculating the given mean velocity in the prototype. The mean velocity in the prototype can be calculated from the flow rate in the prototype divided by the cross-sectional area in the prototype.

$$\frac{Q_p}{A_p} = \frac{Q_p}{\pi r_p^2} = \frac{2\,\frac{\text{cm}^3}{\text{s}}\,\frac{\text{m}^3}{10^6\,\text{cm}^3}}{\pi(2.5/1000)^2\,\text{m}^2} = 0.1018\,\frac{\text{m}}{\text{s}} = 10.2\,\frac{\text{cm}}{\text{s}}$$

Next, recognize that for Reynolds number similarity, the following equation must be true, where the subscript P represent the Reynolds number in the prototype and the subscript m represents the Reynolds number in the model.

$$\left(\frac{\rho VD}{\mu}\right)_p = \left(\frac{\rho VD}{\mu}\right)_m$$

$$\overline{V}_m = \frac{\mu_m}{\mu_p}\frac{\rho_p}{\rho_m}\frac{D_p}{D_m}\overline{V}_p = \frac{1}{4}\frac{1060}{1000}\frac{5}{50}\overline{V}_p = 0.0265\,\overline{V}_p$$

$$\overline{V}_m = 0.0265\,\overline{V}_p = 10.18\text{ cm/s} \times 0.0265 = 0.270\,\frac{\text{cm}}{\text{s}}$$

The velocity in the model should be 0.27 cm/s to achieve Reynolds number similarity to the prototype. This means that we need a water velocity that is 10 times faster than the blood velocity in the venous valve, because we used a model that was 10 time larger.

9.9 Modeling Example 3

The velocity V of an erythrocyte settling slowly in plasma can be expressed as a function of diameter d, thickness t, plasma viscosity μ_p, erythrocyte density ρ_e, plasma density ρ_p, and gravity g.

Written in mathematical formulation, this equals to

$$V_s = f(d, t, \mu, \rho_e, \rho_p, g).$$

Determine a suitable set of Π terms for expressing this relationship. First, write the dimensions of each variable in the equation above.

$$V_s = L/T \quad d = L \quad t = L$$

$$\mu = FT/L^2 \quad \rho_e = \rho_p = M/L^3 \quad g = L/T^2$$

Next, find the number of Pi terms:

$$n = k - r = 7 - 3 = 4 \text{ required Pi terms}$$

The Pi terms could be:

$$\pi_1 = t/d \qquad \pi_3 = \rho_p V_s d/\mu \text{ (Reynolds number)}$$

$$\pi_2 = \rho_e/\rho_p \qquad \pi_4 = V_s^2/gd \text{ (Froude number)}^2$$

Once the important relationships have been determined, it would be possible, for example, to plot the Froude number as a function of Reynolds number for a given thickness/diameter ratio and for a given erythrocyte versus plasma ratio. As long as the Pi terms in the model are equal to the Pi terms in the prototype, that plot would be equally valid for both prototype and model. Figure 9.4 shows a plot of the Reynolds number versus the Froude number for a density of plasma of

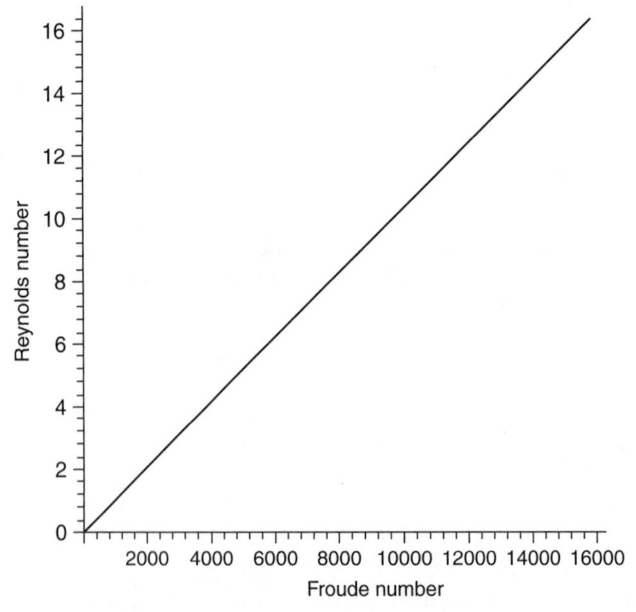

Figure 9.4 The relationship of Reynolds number versus Froude number for the erythrocyte settling problem.

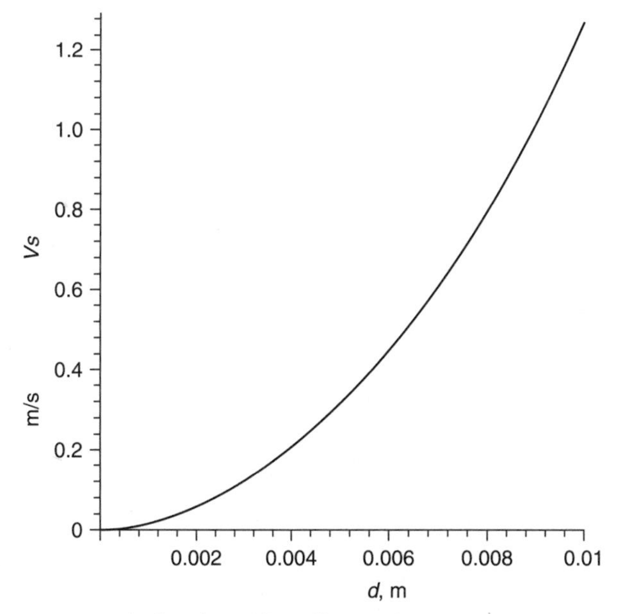

Figure 9.5 A plot of particle sedimentation versus diameter.

1000 kg/m^3 and a density of erythrocytes of 1200 kg/m^3, a viscosity of 0.0035 Ns/m^2, $g = 9.81$ m/s^2, and for a thickness to diameter ratio of 25. Note that this plot has no units and as long as the Pi terms are equal in the model and the prototype, the plot is valid.

Figure 9.5 shows the erythrocyte settling rate as a function of cell diameter for a density of plasma of 1000 kg/m^3 and a density of erythrocytes of 1200 kg/m^3, a viscosity of 0.0035 Ns/m^2, $g = 9.81$ m/s^2, and for a diameter to thickness ratio of 25. Note that this plot has units associated and is thus only valid for this set of parameters.

Bibliography

Munson BR, Young DF, Okiishi TH. *Fundamentals of fluid mechanics*. New York. Wiley; 1994.

Szabo G, Soans D, Graf A, Beller C, Waite L, Hagl S. A new computer model of mitral valve hemodynamics during ventricular filling, *Eur J Cardiothorac Surg*. 2004;26:239–247.

Franck C, Waite L. Mathematical model of a variable aperture mitral valve, *Biomed Sci Instrum*. 2002;38:327–331.

Chapter

10

Lumped Parameter Mathematical Models

10.1 Introduction

Since the pioneering work of Otto Frank in 1899, there have been many types of mathematical models of blood flow. The aim of these models is a better understanding of the biofluid mechanics in cardiovascular systems. Mathematical computer models aim to facilitate the understanding of the cardiovascular system in an expensive and noninvasive way. One example of the motivation for such a model occurred when Raines et al. in 1972 observed that patients with severe vascular disease have pressure waveforms that are markedly different from those in healthy persons. By developing a model that would predict which changes in parameters affect those pressure waveforms in what way, the scientists might provide a means for diagnosing vascular disease before it becomes severe.

Lumped-parameter models are in common use for studying the factors that affect pressure and flow waveforms. A lumped-parameter model is one in which the continuous variation of the system's state variables in space is represented by a finite number of variables, defined at special points called nodes. The model would be less computationally expensive, with a correspondingly lower spatial resolution, while still providing useful information at important points within the model. Lumped-parameter models are good for helping to study the relationship of cardiac output to peripheral loads, for example, but because of the finite number of lumped elements, they cannot model the higher spatial-resolution aspects of the system without adding many, many more elements.

Chapter 10 will begin by addressing the general electrical analog model of blood flow, based on electrical transmission line equations.

Later in Chap. 10 a specific model of flow through a human mitral valve will be presented in some detail.

10.2 Electrical Analog Model of Flow in a Tube

In Chap. 7, Sec. 7.7, a solution was developed, which was published by Greenfield and Fry in 1965, that shows the relationship between flow and pressure for axisymmetric, uniform, fully developed, horizontal, Newtonian, pulsatile flow. The Fry solution is particularly useful when modeling the relationship of the pressure gradient in a tube or blood vessel to the instantaneous blood velocity. In Chap. 8 we used the Fry solution to develop an electrical analog of a typical pressure-measuring catheter. Using well-known solutions to *RLC* circuits, it became possible to characterize parameters like the natural frequency and dimensionless damping ratio of the system.

I describe, here in Chapter 10, an alternative development of the modeling of flow through any vessel or tube based on that analog. Let the subscript *V* represent properties of the blood vessel that we are modeling and the subscript *L* represent properties of the terminating load. In Chap. 7, the terminating load was a transducer, but in the more general case of the isolated blood vessel within a cardiovascular system, the terminating load represents the effects of a group of distal blood vessels, often capillaries.

Figure 10.1 shows the electrical schematic of a circuit representing a length of artery, terminating in a capillary bed. The values for resistance, capacitance, and inductance for each resistor, capacitor, and inductor, respectively, are calculated from the blood vessel properties on a per unit length basis.

Recall that the hydrodynamic resistance of a blood vessel depends on its radius and length as well as the viscosity of the fluid flowing in the

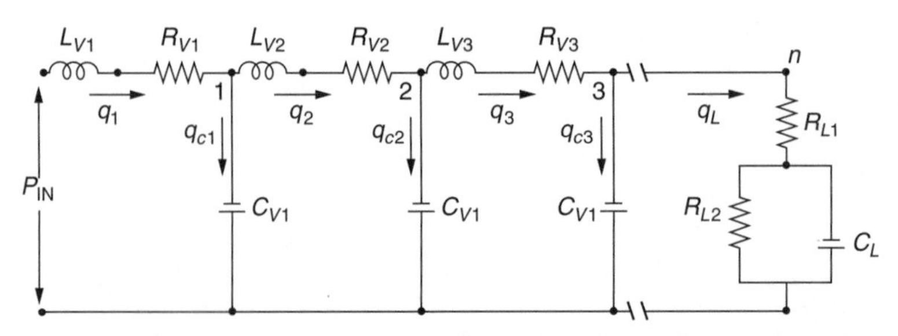

Figure 10.1 The electrical schematic of a model of blood flowing through a vessel or tube with the *V* transcript representing the characteristics of the vessel and *L* representing the terminal load.

tube. For this type of discretized model, the resistance in each discrete resistor can be written as

$$R_V = \ell R_{\text{viscous}} = C_\nu \frac{8\mu\ell}{\pi r^4} = \frac{\text{Ns}}{\text{m}^5} \tag{10.1}$$

The inductance in the vessel for each discrete inductor element becomes

$$L_V = \ell L = C_u \frac{\rho\ell}{\pi r^2} \frac{\text{Ns}^2}{\text{m}^5} \tag{10.2}$$

The capacitance of the vessel is the compliance of the vessel, or dA/dP, and depends on the pressure at the point where the capacitor is located.

$$C_V = \frac{q}{\frac{dP}{dt}} \frac{\text{m}^5}{\text{N}} \tag{10.3}$$

Note that the resistance, inductance, and capacitance of each vessel segment can vary, for example, in a tapering vessel.

10.2.1 Nodes and the equations at each node

In electrical circuit analysis, a node is defined as a point in a circuit where two or more circuit elements join. In Fig. 10.1 we will take the combination of a resistor, inductor, and capacitor as being a single element of our model, and number the nodes at the center of each element at the points where the resistor, capacitor, and inductor are joined. The nodes in Fig. 10.1 are labeled 1, 2, 3,. . ., n.

In the model, the input flow to the first element is labeled q_1. The input pressure is labeled P_{in}. The flow leaving node 1 toward the capacitor is labeled q_{c1}. The pressure at node 1 is P_1. Note that the model does not contain any information about the pressure at points between the input point and node 1.

Considering node 1, we can now write an equation for the pressure drop across the first resistor and inductor as shown in Eq. (10.4) as the first-order differential equation .

$$P_{\text{IN}} - P_1 = q_1 R_{V1} + L_{V1} \frac{dq_1}{dt} \tag{10.4}$$

In Eq. (10.4), P_{IN} represents the input pressure to the vessel. P_1 is the pressure at node 1 and q is the flow through the vessel. The time rate of change of flow is designated dq/dt. Note that L_{V1} and R_{V1} can vary from node to node, and can even vary with pressure, which would cause our model to be nonlinear.

A second differential equation, which is also first order, can be written at node 1. This equation describes the flow into the capacitor at node 1. The flow q_1 is the vessel compliance at this point, C_1 multiplied by dP/dt as shown in Eq. (10.5) . Conceptually, at a specific point in time, it may be useful at this point to think about pressure at each of the nodes, $P_1, P_2, \ldots, P_n$, as being the independent variables and the flows, $q_1, q_2, \ldots, q_n$, as being the dependent variables, which depend on pressure. Mathematically, the P's and q's are dependent on time and location and time is the only independent variable. We will use the model to try to understand the relationship between flow and pressure in the vessel.

$$q_{C1} = C_{V1}\frac{dP_1}{dt} \tag{10.5}$$

This system of two first-order differential equations can be written once for each node in the model. At the end, we will end up with a system of $2n$ first-order, ordinary differential equations, with two first-order equations written for each node: node 1 through node n. Although a large model can be computationally complex, the equations for a single node are relatively straightforward.

If we repeat the two equations for node 2 we end up with the following:

$$P_1 - P_2 = q_2 R_{V2} + L_{V2}\frac{dq_2}{dt} \tag{10.6}$$

$$q_{C2} = C_{V2}\frac{dP_2}{dt} \tag{10.7}$$

The general equations for a general node i, between node 2 and node n will be

$$P_{(i-1)} - P_i = q_i R_{Vi} + L_{Vi}\frac{dq_i}{dt} \tag{10.8}$$

$$q_{Ci} = C_{Vi}\frac{dP_i}{dt} \tag{10.9}$$

10.2.2 Terminal load

If we are modeling the aorta, for example, it is possible to divide the vessel into segments and write a set of equations for each finite segment. If we continue to add the downstream details of every branch of the aorta then the model will become larger and larger, more and more complicated, and more and more computationally expensive. Finally, it becomes impractical to individually model each one of the tens of billions of capillaries in the circulatory system. Instead, a more practical solution is that we will lump together all of the elements downstream of, or distal to, the main vessel that we are trying to understand with our model.

Because the capillaries are primarily resistance vessels we could begin estimating a load resistance that is based on the pressure at node n, and the flow moving through the entire capillary bed. The total, terminal load resistance, R_T is equal to the pressure at node n divide by the total flow q_n as shown in Eq. (10.10).

$$R_T = \frac{P_n}{q_n} \tag{10.10}$$

Although R_T would be a good first-order estimate of the terminal load in our model, we also know from empirical evidence that the capillary bed does not act as a pure resistance element. If we ended the model of the aorta, for example, in a single resistance we would find that pressure waves are reflected proximally because of a mismatch in impedance. In many cases, we would predict standing pressure waves by examining the model, where we know from the empirical data that no standing pressure waves truly exist. That is to say, the pressure along the vessel will not show a steady pressure gradient as one might expect, but a time varying pressure gradient that looks like periodic noise when one plots pressure gradient versus distance along the axis of the vessel. In fact, the vessels downstream of our model that make up the terminal load also exhibit a capacitive effect.

It has been suggested that a terminal load for our model could be estimated as a resistor in series with a second resistor parallel to a capacitor as shown in Fig. 10.2. For steady flow, or at very low frequencies, the total load impedance is equal to the total load resistance, which in this case is the sum of the two resistors. The total terminal resistance R_T is equal to R_{L1} plus R_{L2}, the sum of the values of the two terminal load resistors.

One method of estimating the capacitance of the load, C_L, is by impedance matching. For practical, biological systems, we would expect the output impedance of our model to match the input impedance of the terminal load, so that wave reflections are minimized. The ratio of R_{L1} and

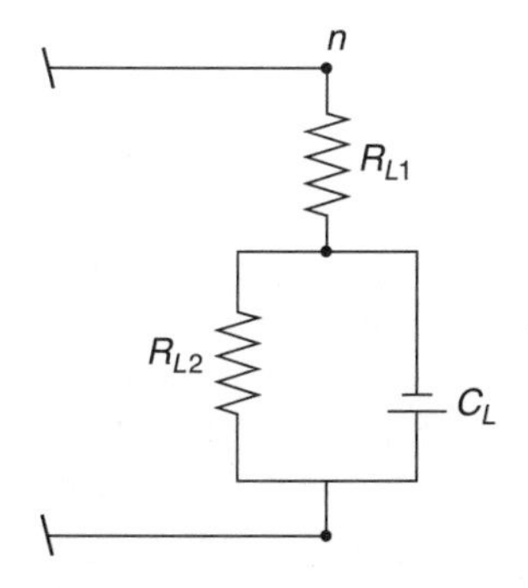

Figure 10.2 The terminal load of the model described in Sec. 10.2.

R_{L2}, as well as the value for capacitance can be chosen to minimize wave reflections and to match the behavior of empirical data as closely as possible.

10.2.3 Summary of the lumped parameter electrical analog model

Many scientists have used similar models to mimic blood flow and pressure conditions in animal circulation as well as in the human circulatory system, with the goal of better understanding of the relationship between the two.

For example, Bauernschmidt et al. (1999) simulated flow hemodynamics during pulsatile extracorporeal perfusion. Control of perfusion is achieved by surgeons, anesthesiologists, and perfusionists making real time decisions. This leads to variations of the perfusion regimens between different hospitals and between different surgical teams in the same hospital. To develop a computer-controlled scheme for the integrated control of extracorporeal circulation, they developed a mathematical model for simulating hemodynamics during pulsatile perfusion. The model was constructed in Matlab using Simulink. They divided the human arterial tree into a 128 segment multibranch structure. Peripheral branches were terminated by resistance terms representing smaller arterioles and capillary beds. Bauernschmidt studied the effects of different perfusion regimens with differing amounts of flow and pulsatility and found the model to be useful for a realistic simulation of different perfusion regimens.

In 2004, L. R. John developed a mathematical model of the human arterial system based on an electrical transmission line analogy. The authors concluded that quantitative variations of blood pressure and flow waveforms along the arterial tree from their model followed clinical trends.

10.3 Modeling of Flow through the Mitral Valve

Assessment of ventricular function and quantification of valve stenosis and mitral regurgitation are important in clinical practice as well as in physiological research (Thomas and Weyman, 1989, 1991; Takeuchi et al., 1991; Thomas et al., 1997; Scalia et al., 1997; Rich et al., 1999; Firstenberg et al., 2001; DeMey et al., 2001; Garcia et al., 2001). Approximately 400,000 patients are diagnosed with congestive heart failure in the United States each year. Elevated diastolic filling pressure in these patients leads to the development of congestive heart failure symptoms (Jae et al., 1997). Noninvasive assessment of diastolic function that does not require the use of intracardiac pressure has been an important goal, and in recent years, Doppler electrocardiography has become the "diagnostic modality of choice" (Garcia et al., 2001) to assess diastolic function.

The goal of the research that created this model was development and validation of a mathematical model of flow through the mitral valve during early diastolic ventricular filling (also sometimes referred to as E-wave filling), as shown in Fig. 10.3. This model may be used to assess diastolic left ventricular function based on Doppler velocity waveforms and cardiac geometry.

A secondary goal of the project was to publish representative values of important input parameters including effective mitral valve area, transvalvular inertial length, blood viscosity, blood density, atrial compliance, ventricular compliance, left ventricular active-relaxiation characteristics, and initial pressure and flow values in the porcine model, which might be of use to other scientists.

By comparing the computer model output values to the measured pressure and flow rate data in pigs, parameters such as atrial compliance, ventricular compliance, and effective area were estimated. This is an inversion of the process used in some earlier papers in which pressure and flow predictions proceed from impedance and compliance data.

The model that was developed, which I will describe in this section, calculates the flow and pressure relationships between the left atrium and the left ventricle through the mitral valve during early ventricular filling prior to atrial systole. I developed this model with the help of some students and in collaboration with cardiac surgeons at the University of Heidelberg Heart Surgery Laboratory. It was first published in the *European Journal of Cardiothoracic Surgery* in September 2004.

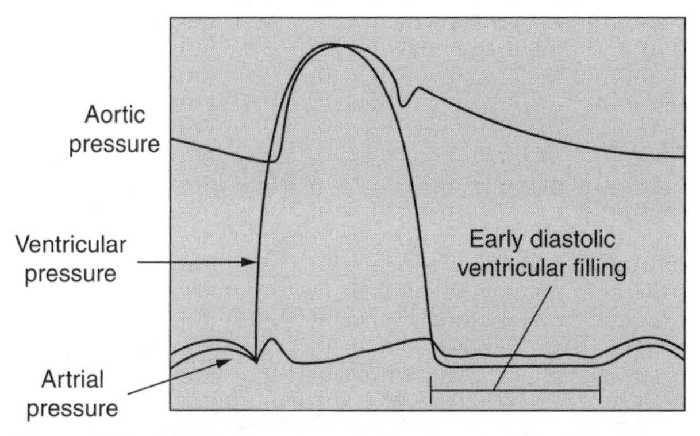

Figure 10.3 A typical pressure waveform in the left atrium, left ventricle, and aorta and the section that is marked shows the period designated as *early diastolic ventricular filling* or *E-wave filling*.

10.3.1 Model description

The model of early ventricular filling, or the E-wave portion of a single heartbeat, begins at the time when pressures are equal in the atrium and the ventricle, i.e., it begins from the instant of mitral valve opening. The model then describes flow and pressures during ventricular filling up to the point of atrial systole. Figure 10.4 shows a schematic of the atrium and ventricle with the mitral valve in-between. A fluid cylinder of length L is shown.

A system of three ordinary differential equations [Eqs. (10.11) to (10.13)] describes the system. For the entire model, I will use the following representations:

P_a = left atrial pressure

q = instantaneous flow rate through the mitral valve

R_c = convective resistance

C_a = atrial compliance

P_v = left ventricular pressure

R_v = viscous resistance

M = inertance term

C_v = ventricular compliance

A = mitral valve effective area

V = average velocity of blood through the mitral valve

R_v, R_c, M, C_a, C_v are described in greater detail in the text following.

The first differential equation describing flow rate through the mitral valve is shown in Eq. (10.11).

$$\frac{dq}{dt} = \frac{(P_a - P_v - R_v q - \operatorname{sgn}(q) R_c q^2)}{M} + V\frac{dA}{dt} \tag{10.11}$$

In the electrical analog, pressures are analogous to voltages in the circuit diagram. Flows are analogous to currents. The convective resistance term is analogous to a resistor, whose resistance is dependent on the current, which flows through it. Convective resistance will be discussed in

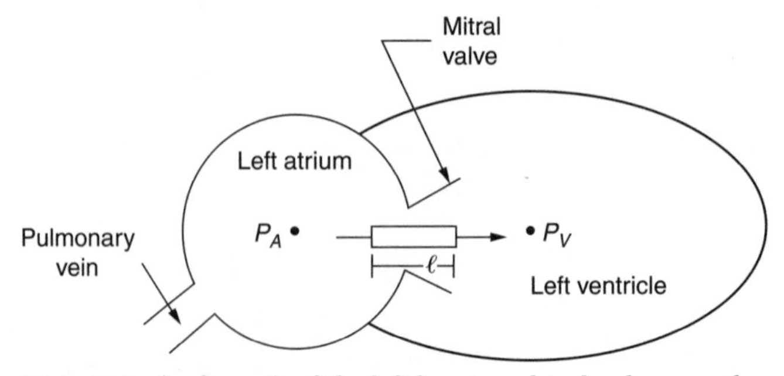

Figure 10.4 A schematic of the left heart used to develop a mathematical model of flow through the mitral valve. P_A is atrial pressure and P_V is ventricular pressure.

more detail in Sec. 10.3.3. The term $V\,dA/dt$ represents the time rate of change of flow rate due to the closing or opening of the mitral valve. If the valve were held completely open, this term would be zero. Section 10.3.4 will describe, in greater detail, the variable-area mitral valve that is used in this model.

The second differential equation is written as follows:

$$\frac{dP_a}{dt} = \frac{-q}{C_a} \tag{10.12}$$

where C_a represents atrial compliance in m^5/N ($m^3/N/m^2$). The pressure inside the atrium is related to its volume through atrial compliance. $C_a = d(\text{volume})/dt$. The volume is also related to flow rate. In one sense, the flow intro the atrium determines its final volume and therefore the final atrial pressure.

Similarly, the third equation describing the system is the equation for change in ventricular pressure with time. The ventricular pressure is similarly related to the ventricular volume and the ventricular compliance, C_v. The change in pressure of the ventricle is also related to "active relaxation." The active relaxation term, resulting from ventricular geometry, is described in more detail in Sec. 10.3.2.

$$\frac{dP_v}{dt} = \frac{q}{C_v} - \text{active relaxation term} \tag{10.13}$$

In all three equations, q represent flow rate in m^3/s, dq/dt represents the time rate of change of flow rate in m^3/s^2, P_a represents left atrial pressure in N/m^2, P_v represents left ventricular pressure in N/m^2, and M represents the inertance term, which is analogous to inductance in the electrical circuit and has the units of kg/m^4.

M in Eq. (10.14) can be calculated as

$$M = \rho l/A \tag{10.14}$$

where ρ represents blood density in kg/m^3, l represents the blood column length through the mitral valve with units of meters, and A represents the mitral valve effective area in m^2.

R_v in Eq. (10.15) represents viscous resistance, which is analogous to resistance in the electrical circuit and has the units kg/m^4s. R_v can be calculated as

$$R_v = 8\rho l\mu/A^2 \tag{10.15}$$

where μ represents viscosity with units of Ns/m^2.

R_c represents convective resistance, which, in an electrical circuit model, would be analogous to a resistor whose variable resistance is dependent on current through the resistor, and has the units kg/m^7. R_c can be calculated as

$$R_c = \rho l / A^2 \tag{10.16}$$

10.3.2 Active ventricular relaxation

Another parameter affecting the final ventricular pressure during filling is active relaxation. To imagine active relaxation, think of an empty plastic 1 L coke bottle. If you squeeze the bottle, the volume decreases, forcing air out of the bottle. Now when you release the bottle because of its geometry the bottle will expand, and if air is not allowed to flow back into the bottle (place your hand over the entrance) the pressure inside it will decrease below atmospheric. At the beginning of diastole, the left ventricle expands in a similar way and we say that active ventricular relaxation is present.

For the purpose of the model, active relaxation can be measured empirically and included in the model as a term that decreases the pressure in the ventricle. When active relaxation occurs, it looks like a step change of pressure in a near critically damped system. Active relaxation parameters include a time constant for the pressure drop off and also a ratio of the maximum atrial pressure (pressure just prior to active relaxation) to minimum atrial pressure.

10.3.3 Meaning of convective resistance

To help understand the meaning of convective resistance, consider the Bernoulli equation while neglecting viscous resistance. The transmitral pressure drop, ΔP, is equal to $1/2\ \rho\ V^2$ where V is the transmitral velocity. R_c represents the convective resistance or resistance to flow associated with acceleration due to spatial changes rather than temporal changes. Blood must speed up as it passes through the mitral valve, because the valve acts like a nozzle on a garden hose. By using a nozzle on a garden hose, you can increase the velocity of the water as it exits the hose without increasing the pressure driving the water. In the same way that water is speeded up when leaving a garden hose using a nozzle, blood must be accelerated as it moves through the mitral valve. That acceleration is the explanation for convective resistance.

10.3.4 Variable area mitral valve model description

Previous lumped parameter models of diastolic filling modeled the mitral valve as a cylinder of a fixed cross-sectional area and length (Thomas and Weyman, 1989, 1992; Faschkampf et al., 1992; Waite et al., 2000). Although, the valve opening is clearly not constant since the valve by nature must open and close, it was possible to make a reasonable first estimate of the flow through the valve by modeling it as instantaneously 100 percent open or 100 percent closed.

In the current version of the model, the valve is modeled with a time varying mitral opening area. One piece of the lumped-parameter model described earlier is a model of the intrinsic and extrinsic characteristics of the valve itself, disregarding the heart chambers. We will call this portion of the model the "variable-area mitral valve model." A complete description of the variable area mitral valve model that was used as a part of this model can be seen in Franck and Waite (2002).

Using a systems approach, the mitral valve aperture can be viewed abstractly as a mechanical system whose behavior is governed by intrinsic dynamics and the forces acting on it. The intrinsic dynamics of the valve aperture are modeled by a second-order linear differential equation. This way of modeling takes into account the mass of the valve cusps, the elasticity of the tissue, and the damping experienced by the valve cusps while it also helps to keep the valve model simple. The equation already contains two of the six parameters the valve model uses: the damping coefficient D and the natural frequency of the mitral valve ω. D describes the amount of damping the valve cusps experience, which mainly depends on blood viscosity and the size of the cusps. The natural frequency ω takes into account the mass of the valve cusps and the elasticity of the tissue.

$$\frac{1}{\omega^2}\ddot{A} + \frac{2D}{\omega}\dot{A} + A = F(t) \tag{10.17}$$

The intrinsic dynamics of the forces acting on the valve, $F(t)$, are represented by the following equation with the parameters explained later in Sec. 10.3.5.

$$F(t) = (A_{\max} - A)\left[K_s(P_a - P_v) + K_d \mathrm{sgn}(V)V^2 + K_a \frac{dV}{dt}\right] \tag{10.18}$$

Equation (10.17) is a second-order, nonlinear differential equation. The variable area A replaces all occurrences of the constant effective mitral valve area in the earlier model.

10.3.5 Variable area mitral valve model parameters

The equation for the variable area introduced six new parameters into the model.

Parameter1. A_{max} is the cross-sectional aperture of the mitral valve when fully opened.

Parameter 2. The Greek character ω represents the natural frequency of the mitral valve in s^{-1}. It is influenced by the mass of the valve cusps and the modulus of elasticity of the tissue.

Parameter 3. D represents the damping coefficient of the mitral valve. The valve cusps have comparatively little mass, but have a large area and they move in blood, a viscous fluid. Therefore, the system will possess significant damping (D>1). The viscosity of the blood and the size of the valve cusps govern the magnitude of the damping coefficient. The damping coefficient has to be chosen so that no overshoot occurs, i.e., the actual mitral valve area must never be greater than the maximum area.

Parameters 4-6 K_s, K_d, and K_a: These "gain factors" represent the sensitivity of the valve to static, dynamic, and acceleration-induced pressure. K_s must be unitless, while K_d has the unit mass/volume and K_a has the unit mass/area.

10.3.6 Solving the system of differential equations

The model described here in Sec. 10.3 uses the Matlab command ode45 to solve the system of ordinary differential equations. The command ode45 is based on an explicit Runge-Kutta formula, the Dormand-Prince pair. It is a one-step solver. That means that when computing y at some time step n, it needs only the solution at the immediately preceding time point, y at time point ($n - 1$). Matlab recommends ode45 as the best function to apply as a "first try" for solving most initial value problems.

10.3.7 Model trials

The model was validated by collecting empirical data on pigs at the University of Heidelberg, Heart Surgery Laboratory (Szabo et al., 2004). When using a mathematical model with many parameters, it becomes an important task to determine which parameters will be considered inputs and which will be outputs. If we have 100 parameters in the model, and 99 are considered inputs, and one is considered the output of the model, we need to decide a priori which parameter is most important to select.

Primary inputs to the model during the Heidelberg study were the diastolic atrial and ventricular pressure curves, a parameter describing left-ventricular active relaxation, mitral valve geometry, blood density, and blood viscosity. For these trials, atrial and ventricular compliance can also be considered model inputs. In other situations, it might be useful to estimate the values for these parameters from the model. The primary output of the model was calculated flow across the mitral valve. The waveforms of experimental and calculated flow across the mitral valve were compared to validate the model.

10.3.8 Results

Table 10.1 shows the various parameter values used for modeling the six porcine trails. The numbers in column 3 represent the data used in the trial shown in Fig. 10.5. In that figure, typical results measured in trial 3 in a porcine model are shown. The figure shows that when the atrial and ventricular pressures are correctly modeled and closely match the experimentally measured pressures, the calculated flow of blood through the mitral valve also matches the experimentally measured value. This also means that the measured stroke volume also compares closely to the stroke volume predicted by the model.

The following parameters were held constant in all simulations; time duration of simulation = 0.137 s, integration step size = 0.0049 s, offset = 0 mmHG, $\mu = 0.035$ dyne s/cm^2, $\rho = 1$ gm/cc. MSV denotes *measured stroke volume* from data and SSV denotes *simulation stroke volume*, both have units cubic centimeter.

Table 10.1 shows the MSV compared to the SSV. The data demonstrate that flow through the mitral valve in the porcine model can be modeled successfully with comparable stroke volumes. By matching flow and

TABLE 10.1 List of the Various Parameter Values Used for Modeling the Six Porcine Trails

	1	2	3	4	5	6	Average
α_a, cm^{-3}	0.26	0.079	0.09	0.05	0.06	0.029	0.095
α_v, cm^{-3}	0.17	0.22	0.21	0.12	0.12	0.16	0.17
l, cm	1.5	2	1.5	1.5	1.5	1.7	1.6
K_a, g/cm^2	0.002	0.002	0.002	0.0005	0.001	0.001	0.0014
K_d, g/cm^3	0.003	0.002	0.0015	0.0012	0.0003	0.0017	0.0016
K_s	0.007	0.002	0.0015	0.0001	0.0009	0.001	0.0021
Max valve aperture, cm^2	2	0.8	1.7	1.7	1.8	0.95	1.5
ω, rad/s	30	30	30	30	30	25	29.2
D, g/s	10	10	10	10	10	5	9.2
MSV, cc	16.6	9.6	11.4	12.8	11.3	4.9	11.1
SSV, cc	10.0	10.1	10.5	14.1	11.6	8.8	10.9

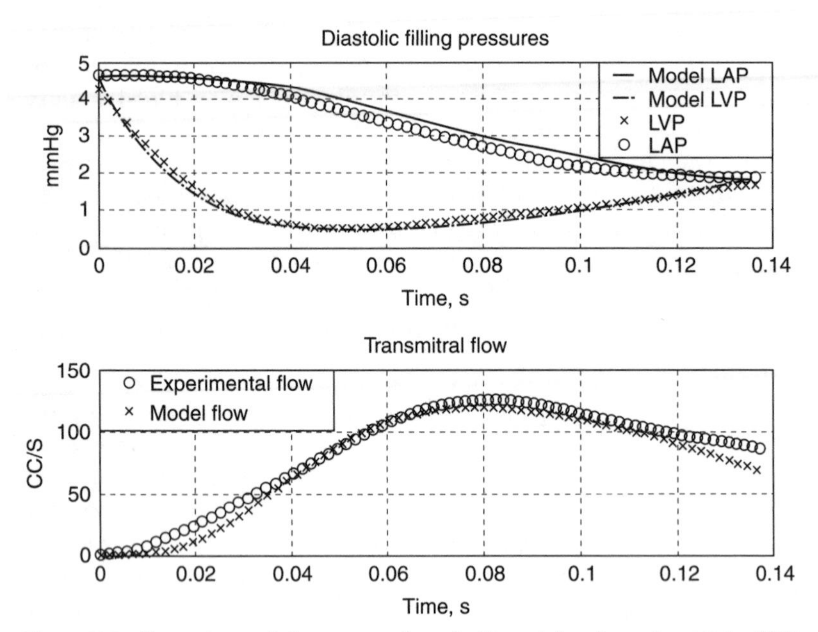

Figure 10.5 Pressures and flow curves from both model and porcine trials. *LVP* denotes left ventricular pressure; *LAP* stands for left atrial pressure. The different parameter values for this particular trial are listed in column 3 in Table 10.1.

pressure waveforms, it was possible to predict an unknown parameter like atrial compliance.

10.4 Summary

Chapter 10 presents an introduction to lumped parameter models and shows a few examples of their uses. Research that leads to a better understanding of the cardiovascular system in humans is important, but also problematic. In general, because of ethical considerations, it is not possible to perform research on humans simply to gain a better understanding of how the body functions. Even if we are able to simulate some aspects of human systems with animal research, that type of research is very expensive and can even be wasteful, and inhumane in the worst of cases. The development of mathematical models that can noninvasively simulate important parameters in human systems is a lofty goal and one that should be embraced.

On the other hand, without proper validation in appropriate animal or human systems, these models are sometimes little better than educated guesses. Chapter 10 introduces a mathematical model of flow that is based on equations similar to those of the transmission line equations in electrical engineering. Examples of published uses of similar models are shown.

Finally, in Chap. 10, I present a more in depth overview of a lumped parameter model of flow through the mitral valve. The model is based on a system of nonlinear ordinary differential equations that are solved in Matlab using the ode45 command. The chapter presents some typical results from a porcine model and shows some promising results from that model.

Approximately 400,000 patients are diagnosed with congestive heart failure in the United States each year. Elevated diastolic filling pressure in these patients leads to the development of congestive heart failure symptoms. Noninvasive assessment of diastolic function that does not require the use of intracardiac pressure has been an important goal. In recent years, Doppler electrocardiography has become the "diagnostic modality of choice" to assess diastolic function. The goal of the model described in Chap. 10 was development and validation of a mathematical model of flow through the mitral valve during early diastolic ventricular filling (also sometimes referred to as E-wave filling), which promises assessment of diastolic left ventricular function using noninvasively collected Doppler waveforms.

Bibliography

Bauernschmitt R, Naujokat E, Mehmanesh H, Schulz S, Vahl CF, Hagl S, Lange R. Mathematical modelling of extracorporeal circulation: simulation of different perfusion regimens, *Perfusion*. 1999 September ;14(5):321–30.

Bauernschmitt R, Schulz S, Schwarzhaupt A, Kiencke U, Vahl CF, Lange R, Hagl S. Simulation of arterial hemodynamics after partial prosthetic replacement of the aorta, *Ann Thorac Surg. 1999 October;68(4):1441–2.*

Fisher J. Comparative study of the hydrodynamic function of six size 19 mm bileaflet heart valves, *Eur J Cardiothorac Surg*. 1995;9(12);692–5 discussion 695–6.

Flachskampf F, Weyman A, Guerrero JL, Thomas JD. Calculation of atrioventricular compliance from the mitral flow profile: analytic and in-vitro study, *J Am Coll Cardiol*. 1992;19:998–1004.

Franck C,Waite L. Mathematical model of a variable aperture mitral valve, *Biomed Sci Instrum*. 2002;38: 327–331.

Franck C. *A lumped variable model for mitral valve aperature during diastolic filling* [M.S. thesis]. Rose-Hulman Institute of Technology, Terre Haute, IN; 2001.

Garcia M, Firstenberg M, Greenberg M, Smedira M, Rodriguez L, Prior D, Thomas J. Estimation of left ventricular operating stiffness from Doppler early filling deceleration time in humans, *Am J Physiol Heart Circ Physiol*. 2001;280(2):H554–561.

Gorlin R, Gorlin SG. Hydraulic formula for calculation of the area of the stenotic mitral valve, other cardiac valves, and central circulatory shunts, *Am Heart J*. 1951;41(1):1–29.

Greenfield JC, Fry DL. Relationship between instantaneous aortic flow and the pressure gradient, *Circ Res*. 1965;17:340–348.

Isaaz K. A theoretical model for the noninvasive assessment of the transmitral pressure-flow relation, *J Biomech*. 1992;25(6):581–90.

Jae KO, Appleton CP, Hatle LK, Nishimura RA, Seward JB, Tajik AJ. The noninvasive assessment of left ventricular diastolic function with two-dimensional and Doppler echocardiography, *J Am Soc Echocardiogr*. 1997;10:246–270.

John LR. Forward electrical transmission line model of the human arterial system, *Med. Biol. Eng. Comput*. 2004 May;42(3):312–21.

Lemmon J, Yoganathan AP. Computational modeling of left heart diastolic function: Examination of ventricular dysfunction, *J Biomech Eng*. 2000;122(4):297–303.

Nudelman S, Manson A, Hall A, Kovacs S. Comparison of diastolic filling models and their fit to transmitral Doppler contours, *Ultrasound Med Biol*. 1995;21(8):989–999.

Raines JK, Jaffrins MY, Shapiro AH. A computer simulation of arterial dynamics in the human leg, *Biomechanics*. 1974;7:77–91.

Reul H, Talukder N, Müller EW. Fluid dynamics model of mitral valve, *J Biomech*. 1981;14(5):361–372.

Rich M, Sethi G, Copeland J. Assessment of left ventricular compliance during heart preservation, *Perfusion*. 1998;13:67–75.

She JD, Yoganathan AP. Three-dimensional computational model of left heart diastolic function with fluid-structure interaction, *J Biomech Eng*. 2000;122(2):109–17.

She J, Li M, Huan L, Yu Y. Dynamic characteristics of prosthetic heart valves, *Med Eng Phys*. 1995;17(4):273–81.

Szabo G, Soans D, Graf A, Beller C, Waite L, Hagl S. A new computer model of mitral valve hemodynamics during ventricular filling, *Eur J Cardiothorac Surg*. 2004;26:239–247.

Takeuchi M, Igarashi Y, Tomimoto S, Odake M, Hayashi T, Tsukamoto T, Hata K, Takaoka H, Fukuzaki H. Single-beat estimation of the slope of the end-systolic pressure-volume relation in the human left ventricle, *Circulation*. 1991;83:202–212.

Thomas JD, Newell J, Choong C, Weyman A. Physical and physiological determinants of transmitral velocity: numerical analysis, *Am J Physiol*. 1991;260:H1718–H1730.

Thomas JD, Weyman AE. Numerical modeling of ventricular filling, *Ann Biomed Eng*. 1992;20:19–39.

Thomas JD, Weyman A. Fluid dynamics model of mitral valve flow: description with in-vitro validation, *J Am Coll Cardiol*. 1989;13:221–33.

Thomas JD, Weyman A. Echocardiographic Doppler Evaluation of Left Ventricular Diastolic Function, *Circulation*. 1991; 84(3):977–990.

Thomas JD, Zhou J, Greenberg N, Bibawy G, Greenberg N, McCarthy P, Vandervoort P. Physical and physiological determinants of pulmonary venous flow: numerical analysis, *Am J Physiol*. 1997;5(2):H2453–2465.

Thomas J. Prognostic value of diastolic filling parameters derived using a novel image processing technique in patients > or = 70 years of age with congestive heart failure, *Am J Cardiol*. 1999;84(1):82–6.

Verdonck P, Segers P, Missault L, Verhoeven R. In-vivo validation of a fluid dynamics model of mitral valve M-mode echocardiogram, *Med Biol Eng Comput*. 1996;34(3):192–198.

Index

Page numbers followed by *t*, *f*, or *n* indicate tables, figures, or notes, respectively.

ABOUT THE AUTHOR

LEE WAITE is Chair of the Department of Applied Biology and Biomedical Engineering, and Director of the Guidant/ Eli Lilly and Co. Applied Life Sciences Research Center, at the Rose-Hulman Institute of Technology in Terre Haute, Indiana. He is also the president of the Rocky Mountain Bioengineering Symposium (RMBS). Held annually since 1964, the RMBS is the longest continually operating biomedical engineering conference in North America.

612.1 Waite, Lee
WAI Biofluid mechanics in...
2006 112650